Books of the Body

BOOKS
of the
BODY

*Anatomical Ritual
and Renaissance Learning*

Andrea Carlino

TRANSLATED BY
JOHN TEDESCHI AND ANNE C. TEDESCHI

THE
UNIVERSITY OF CHICAGO PRESS
Chicago and London

Andrea Carlino is a member of the faculty at the Institut Romand d'Histoire de la Médecine, University of Geneva, Switzerland. He is the author of *Paper Bodies: A Catalogue of Anatomical Fugitive Sheets in the Age of Printing and Dissecting* and coeditor of *Corps à vif: Art et anatomie.* John Tedeschi and Anne C. Tedeschi have translated seven books from the Italian, including *The Cheese and the Worms,* by Carlo Ginzburg.

Originally published as *La fabbrica del corpo: Libri e dissezione nel Rinascimento,* by Andrea Carlino. © 1994 by Giulio Einaudi editore s. p. a., Turin.

The following organizations have generously granted permission to reproduce illustrations: figs. 1, 18, and 19, Electa Archive, Milan; figs. 2, 12, 13, 16, and 17, Bibliothéque Nationale de France; fig. 6, Istituti Ortopedici Rizzoli, Bologna; figs. 7, 20, 22, 23, 24, 27, and 32, Biblioteca Angelica and the Ministerio del Tesoro, Italia; figs. 8 and 14, Biblioteca Marucelliana, Florence; fig. 21, Museo Civico, Bassano del Grappa; fig. 25, Glasgow University Library, Department of Special Collections; fig. 26, Mauritshuis, The Hague; fig. 30, Kunsthistorisches Museum, Vienna; fig. 31, The Royal Collection, Her Majesty Queen Elizabeth II; figs. 33–38, Wellcome Institute Library, London.

The University of Chicago Press, Chicago 60637
The University of Chicago Press, Ltd., London
© 1999 by The University of Chicago
All rights reserved. Published 1999
08 07 06 05 04 03 02 01 00 99   1 2 3 4 5
ISBN: 0-226-09287-9 (cloth)

Library of Congress Cataloging-in-Publication Data

Carlino, Andrea, 1960–
[Fabbrica del corpo.   English]
Books of the body : anatomical ritual and renaissance learning / Andrea Carlino ; translated by John Tedeschi and Anne C. Tedeschi.
p.   cm.
Includes bibliographical references and index.
ISBN 0-226-09287-9 (cloth : alk paper)
1. Human dissection—History—16th century.   2. Human anatomy—History—16th century.   3. Renaissance.   I. Title.
QM33.4.C3613   1999
611'.009'031—dc21   99-25338
CIP

♾ The paper used in this publication meets the minimum requirements of the American National Standard for Information Sciences—Permanence of Paper for Printed Library Materials, ANSI Z39.48-1992.

Plus ça change, plus c'est la même chose.
  A. Karr, *En fumant*

. . . nec est ulla supra terras adeo rabiosa belua cui non imago sua sancta sit.
  Quintilian, *Declamationes*, XII: 27

Contents

Illustrations

Acknowledgments

This book is the result of a series of conversations with Roger Chartier, Armando Petrucci, and Joseph Rykwert. Together with Adriano Prosperi they have guided and counseled me through every step of the research and writing process.

During that time I received valuable suggestions and engaged in informative discussions with Mario Agrimi, Leonard Barkan, Michele Bernardini, Massimo Bray, Andrea Viana Daher, Anne Engel, Carla Frova, Carlo Ginzburg, Giovanni Levi, Sergio Luzzato, Agata Moretti, Josè Maria Perceval, Jackie Pigeaud, Marie-Domitille Porcheron, Jacques Revel, Franco Rizzi, Elisa Romano, Aldo Schiavone, Giorgio Stabile, Francesco Trevisani, Mario Vegetti, Pippo Vitiello, Corrado Vivanti, and my companions at the *Scuola Superiore di Studi Storici* at the University of San Marino.

I presented parts of the present research at the seminar directed by Roger Chartier at the École des Hautes Études en Sciences Sociales (Paris) and at two conferences: *Les Corps à la Renaissance* (Tours, 1987) and *Il mestiere di leggere, il piacere di leggere* (San Giovanni Valdarno, 1987), as well as at the Lindisfarne conference, "Conversations on the Historicity of the Body" (Columbia, MD, 1992). A preliminary version of chapter 1 appeared as an article titled "The Book, the Body, the Scalpel," in *RES* 16 (1988): 32–50.

In addition to those people already mentioned, I should like to thank the personnel of the following archives and libraries: Archivio di Stato, Rome; Archivio Segreto Vaticano; Biblioteca Angelica; Biblioteca Apostolica Vaticana; Biblioteca Casanatense; Bibliothèque de l'École Française de Rome; Bibliothèque de France, Paris; the British Library; the Warburg Institute Library; and the Wellcome Library.

Finally, I should like to thank my mother for her affectionate concern and to express my special gratitude to Marina Engel, who during these years helped me greatly by listening to me, advising me, and . . . putting up with me.

This book in the present English edition is substantially the same as the Italian version entitled *La fabbrica del corpo* published in 1994. I have introduced a few slight changes to facilitate its reading and corrected a number of errors. More importantly, I have added a bibliography, epilogue, and index, omitted in the Italian edition. The publication of this work by the University of Chicago Press and its accompanying revision have been made possible thanks to the valuable suggestions of two anonymous readers, to the counsel of Nancy Siraisi, and, above all, to the patience and enthusiasm of Susan Abrams. A final, special thanks goes to all those who have collaborated to make this edition possible: first of all Anne and John Tedeschi, historians, bibliophiles, and meticulous translators, whom I had the good fortune to encounter for this difficult and challenging assignment; Ruth Barzel and Erik Carlson, who had the editing responsibilities; and, finally, my mother-in-law, Anne Engel, who skillfully counseled me and participated through the entire process of reading the manuscript and the proofs. To all go my warm and sincere thanks.

Geneva, June 1999

ABBREVIATIONS

ARISTOTLE

GA	*De generatione animalium*
HA	*Historia animalium*
PA	*De partibus animalium*
ASB	Archivio di Stato, Bologna
ASR	Archivio di Stato, Rome
ASV	Archivio Segreto Vaticano
AS Ven.	Archivio di Stato, Venice
BAV	Biblioteca Apostolica Vaticana
BN	Biblioteca Nazionale Vittorio Emanuele II, Rome
CGM	*Corpus Medicorum Graecorum*, Berlin-Leipzig, 1908 ff.

GALEN

De anat.	*De anatomicis administrationibus*
De usu	*De usu partium corporis humani*
De sectis	*De sectis ad eos qui introducuntur*
HC	*L. Hain, Repertorium bibliographicum in quo libri omnes ab arte typographica inventa usque ad anno MD. Milan, 1966; W. A. Copinger, Supplement to Hain's Repertorium Bibliographicum. Milan, 1950.*
IGI	*Indice generale degli incunaboli delle biblioteche d'Italia. Rome, 1943–81.*

K	C. G. Kühn, *Claudii Galeni Opera Omnia*, Leipzig, 1821–33
Littré	*Oeuvres complètes d'Hippocrate,* ed. E. Littré. Paris, 1839–61.
PG	Migne. *Patrologia Graeca*
PL	Migne. *Patrologia Latina*
RE	A. Pauly. *Real-Encyclopädie der Klassischen Altertumswissenschaft.* Ed. G. Wissowa. Stuttgart, 1894 ff.
Statuta 1531	*Bulla de Protomedici et Collegii Medicorum Urbis iurisdictione et facultatibus.* Rome, 1531.
Statuta 1676	*Statuta collegii DD. Almae Urbis Medicorum, ex antiquis Romanorum Pontificum Bullis congesta, et hactenus per Sedem Apostolica recognita et innovata. Mox ab Urbano Octavo confirmata, eorundem Statutorum in Apostolicis litteris inserctione corroborata. Denum a S. D. N. Clemente X firmius consolidata et novis auctoriis amplificata.* Rome: Ex Typ. Rev. Cam. Apost., 1676.
Stat. Gymn. Pat.	*Statuta Almae Universitatis DD. Philosophorum et Medicorum Cognomento Artistarum Patavini Gymnasii. Denuo correcta et emendata et nonnullis apostillis scitu digni aucta.* Padua: Rosatum Bidellum Gen. Almae Univ., 1607.

For the works of Galen, I have always cited from C. G. Kühn, *Claudii Galeni Opera Omnia,* 20 vols., Leipzig, 1821–33. In particular, for the *De anatomicis administrationibus,* even though I have always referred to Kühn, I have also used the Italian translation by Ivan Garofolo: Galeno, *Procedimenti anatomici,* 3 vols., Milan, 1991. For the citations from Aristotle, *De partibus animalium* and *De generatione animalium,* I have used the Italian translation *Parti degli animali* and *Riproduzione degli animali* of Mario Vegetti and Diego Lanza (Bari, 1973).

INTRODUCTION

In 1543 Andreas Vesalius published his *De humani corporis fabrica libri septem* in the Basel printing shop of Johannes Oporinus. The event marked a turning point in the history of anatomy, since he proposed in this work a revision of human morphology as it had been set out in the classical texts and taught in every European university. The revision had been made possible, the author informs us, by repeated observation of cadavers that he had dissected with his own hands. This approach to learning about the parts of the human body was not taken for granted until a few years before the *Fabrica* was written.

The book's greatest merit beyond the correction of the many errors in Galenic anatomy lay in its proposal of a research method that affirmed the educational potential of dissection on one hand and, on the other, provided a tool for an empirical verification of that anatomical knowledge that had been passed down from older writings; dissection was the only possible guide to the trustworthy description of the parts of the human body. Older iconographic representations of the anatomy lesson showed dissection practiced in a way that subordinated observation to description. In opposition to this tradition Vesalius introduced a method that was both didactic and investigative. According to this new method the body dictated the text, with an emphasis on the decisive contribution of the visual to both anatomical teaching and research. With this aim in mind, and following the dictates of Greek and medieval biologists and physicians who sought to chart and reproduce the human body in images, Vesalius provided the *Fabrica* with a series of magnificent illustrations that make the book one of the masterpieces of sixteenth-century printing. They translate the renewed use of dissection into printed images.

The first sporadic evidence of anatomical dissection in the Christian West turns up in Bologna between the end of the thirteenth and the early fourteenth century. From the first decades of the *Quattrocento* onward dissection was part of university educational practice. A special chapter specifically dedicated to anatomy was introduced into the academic statutes of Bologna (1405) and Padua (1465). It established that a demonstration, defined as a "public anatomy" (*anatomia publica*) could be performed annually before an audience of teachers and students on cadavers provided by judicial authorities. We can assume from these norms that dissections could only take place for didactic purposes, that they were for the most part entrusted to barbers and surgeons (traditionally the repositories of the techniques required), and that they were intended to "illustrate" the books read by medical students and commented on by the teachers, thus providing a visual aid to facilitate the memorization of these texts. During the entire sixteenth century such writings consisted of Galen's surviving works and of a series of *compendia*—manuals and treatises inspired by them—of which the most popular was Mondino dei Liuzzi's *Anatomia,* written in 1316.

In opposition to what seemed a blind adherence to classical authorities, a number of physician-anatomists active in Italy between the end of the fifteenth and the first decades of the sixteenth century, such as Alessandro Benedetti, Jacopo Berengario da Carpi, and Niccolò Massa, had noted in their writings that the observation of cadavers provided extraordinary possibilities for research as well as for education. They claimed that they themselves had performed countless dissections, in violation of academic ritual. Galenic anatomy was not affected in the short term, however.

This brief chronological survey serves as my point of departure. Dissection provided an immediate means of learning anatomy by approaching the human body—the source of this knowledge—directly. It constituted, above all, an investigative tool, a technique for unveiling the secrets concealed by the exterior body covering without the mediation of texts. This was a tool that could resolve the contradictions and inconsistencies that these texts sometimes contained. The moment that dissection began to be permitted anatomy could have been rewritten and freed from the authority of Galen and the monopoly of the Galenists. This did not happen, perhaps because between the fourteenth and sixteenth centuries cadavers were cut up at the universities for exclusively educational purposes. The question then becomes: why did two centuries have to pass before there could be a change in the application of human dissection? Was direct observation not (as good sense and the ancient authorities suggest) the principal instrument for acquiring a general knowledge of nature and of the human body in particular? What sort of resistance was there to the effective practice of dissection and to the consequent revision of the anatomical paradigm Galen had founded?

In attempting to answer the questions raised by anatomy's uneven course I have had to reverse the approach usually adopted in the history of science: instead of examining the evolution of fifteenth- and sixteenth-century anatomy, the innovations introduced by the *Fabrica* of Vesalius and the discoveries of the anatomists (which have been discussed elsewhere), I shall try to uncover the reasons for this inertia and the nature of the obstacles blocking its progress. I shall concentrate on a number of factors inherent in anatomical knowledge and in the practice of dissection that, given an exclusively historicist approach, could have obscured, in particular, the decisive role played in this story by psychological elements (the general attitude of physicians who continued to believe and accept the authority of the ancients); by sociological issues (the socioprofessional shift in the wake of the anatomical paradigm), and, especially, by anthropological problems (the revulsion generated by contact with cadavers). This means setting aside, at least partly, the history of anatomy in a narrow sense, and investigating, instead, despite the chronological irregularities and discontinuities, the larger questions that affect the relationship of this science to other cultural determinants.

It seems clear that the procurement and cutting open of cadavers for scientific (thus profane) purposes, and the inevitable delay in the burial of the dead that followed, were considered religiously and anthropologically dangerous acts. To illustrate the risks implied in the performance of these acts by the extrascientific elements influencing the acquisition of knowledge about the human body, I have analyzed as exhaustively and as closely as possible the anatomical practices in a single, significant context: Rome. This city was both the capital of the Christian world and the seat of a university where dissection was regularly carried out and where academic activity was shrewdly filtered through the religious authorities and controlled by political and judicial institutions. The norms governing the practice of public anatomies in the *Studium Urbis* in the sixteenth century, the criteria followed for the selection of the subjects of dissection, and the procedures and acts before, during, and after the desecration of their remains, emerge as a series of prudent strategies implemented to limit the circumstances under which anatomical demonstration could be legitimately allowed. Public anatomy in Rome, as in other Italian university seats, was a rigid academic ceremony. The ritual surrounding dissection seems to have served to domesticate and curb the anthropological risks of the operation.

Along with these nonscientific limitations, the study of classical anatomy and the way in which this knowledge was transmitted to the Renaissance inhibited the development of dissection decisively and led to stagnation in research. Renaissance anatomists, including Vesalius himself, both in their teaching and in their empirical investigations, acknowledged that they set out with the same epistemological presuppositions, that they posed the same order of questions and used the same techniques as those scholars of antiquity who had defined what could be called the anatomical canon. My research then took me along another, wider path. Following indications suggested in those Renaissance writings I traveled back to the beginnings of the history of anatomy in order to identify those characteristics of the discipline and those categories on which both the ancient biologists as well as fifteenth- and sixteenth-century physicians based their reflections on anatomy and on the practice of dissection. I have used a type of research I am tempted to call archeological. Taking my cue from the statements and the acts of Renaissance anatomists, I have tried to understand and to define the regulations and imperatives of the practices connected to anatomical science, and to retrace the genealogical thread and the mechanisms of transmission over the centuries of its epistemological foundations.

Ancient anatomy, as Ludwig Edelstein demonstrated in a famous article published in the 1930s, has been an area of investigation for both the physician as well as, and especially for, the natural philosopher.[1] The twofold character of anatomical discourse was even perceived by Galen when he wrote in the *De anatomicis administrationibus* that there are applications for anatomy "that are more useful for philosophers than for physicians."[2] In fact, the discipline was of distinctly limited use for the practicing physician of that time. It served the surgeon to some degree and could occasionally furnish certain information on the etiology of obscure ailments, but in the context of a prognostic, diagnostic, and therapeutic methodology entirely based on the theory of humors it had no clinical application. Galen's interest in anatomy was justified by his idea of medicine, a medicine, as Mario Vegetti has written, with a "high profile" that had much in common with philosophy, though he accepted its limited

1. L. Edelstein, "The History of Anatomy in Antiquity," in Edelstein's *Ancient Medicine,* ed. O. Temkin and C. L. Temkin (Baltimore, 1967), pp. 247–301; originally "Die Geschichte der Sektion in der Antike," in *Quellen und Studien zur Geschichte der Naturwissenschaften und der Medizin* 3, no. 2, [1932–33]).

2. Galen, *De anat.,* K II 286 ff. Cf. chapter 3, pt. 1.

applicability and its clinical inefficiency.[3] Galen believed that anatomy provided the physician with the essential knowledge that allowed him to proceed rationally but that it was not an instrument for saving human lives.[4]

This view of anatomy persisted into the Renaissance, and the treatise writers of the time were of the opinion that the discipline responded not only to the exigencies of medicine but also to the demands of natural philosophy. Such views about the twofold utility of anatomy found in both ancient and Renaissance texts obviously depended on the fact that much of the knowledge acquired about the human body, especially about its internal organs, could not be applied in any significant way within the available clinical paradigm, which had remained basically unchanged for over a millennium. Of course, a knowledge of myology, osteology, and the course of the veins and of the main arteries was indispensable to both the surgeon and the healer. But, as Galen himself lamented, of what immediate utility to etiology, prognosis, diagnosis, and therapy could have been, for example, a knowledge of the muscles of the tongue or of the muscles, the nerves, the subsidiary arteries, and the veins that run through the heart and the viscera, if it was impossible to act upon them in any way?[5] This was "excess" knowledge, knowledge that pertained to a discipline that pursued its objectives regardless of clinical benefit. The descriptions of the parts of the body offered by the writings of the ancient authorities, especially by Galen, provided a more than exhaustive base upon which physicians could be educated and could practice. These ancient and Renaissance texts, by implication, rejected, in the field of medicine, the notion that anatomical research and thus dissection could have any sort of investigative utility. In fact, as Celsus, Rufus, and Galen had already made plain, recourse to dissection was justified primarily for teaching, since it allowed students to memorize the parts of the body—their dimensions, position, form, and arrangement—on the basis of the descriptions furnished by the texts.

3. M. Vegetti, "Modelli di medicina in Galeno," in *Galen: Problems and Prospects,* ed. V. Nutton (London, 1981), reprinted in M. Vegetti, *Tra Edipo ed Euclide: Forme del sapere antico* (Milan, 1983), pp. 113–37.

4. Galen often regretted the inefficacy of anatomical knowledge in clinical practice. For example, " . . . we are obliged to investigate the utility of all the parts, even if this does not contribute at all to the diagnosis of the disorders or to the prognosis of their sequel"; and also " . . . the discussion [over anatomy] has been extended to things that have no utility at all for therapy, prognosis or for the diagnosis of disorders" (*De usu,* XVIII, 1 and 2, K IV 351 and 364–65). And, in Galen, for medicine in general, anatomy appears as "the decisive guarantee of a high epistemological level and the possibility of a demonstrative organization in the Euclidean manner" (M. Vegetti, "Modelli," p. 117).

5. Galen, *De anat.,* K II 284 and 287.

The fragile and ambiguous epistemological status of anatomy that the classical tradition bequeathed to Renaissance medicine constitutes, in my opinion, a fundamental, though neglected, aspect of the history of this discipline. It may be a plausible explanation for the late establishment of academic chairs in anatomy and a reason as to why the works of Vesalius and his followers, which were read, studied, and appreciated by later writers, did not enjoy immediate success in Europe as university textbooks. It also justified, even after Vesalius's violent attacks, the persistence of the practice of dissection, carried out according to the formalized ritual of the public lesson, for strictly didactic ends and as a visual aid for texts. Since, even in the sixteenth century, the advantages of anatomical knowledge were still not recognizable and demonstrable, there was no epistemological justification for an alternative use of dissection: research beyond the boundaries imposed by Galenism remained the prerogative of the intellectual curiosity of the few. Furthermore, a major revision of the Galenic anatomical paradigm, on which the teachings of all the Renaissance medical schools in the Christian West were based, would have thrown into question not only the recognized value of an ancient authority, but also the power and the professional and academic credibility of all those, physicians and teachers, who had embraced it.

The limited opportunities for using anatomical knowledge in healing left a lot of room, moreover, for the persistence through the centuries of an unease (which took different forms in different contexts) generated by the handling and profanation of the remains of the dead. The acts connected with anatomical practice required a strong and recognizable motivation if they were to be made legitimate. It was not possible to consider the cutting open of a cadaver as an anthropologically innocuous deed or to think of delayed burial as religiously irrelevant. In antiquity and in the Middle Ages the voices of the opponents of dissection based their arguments on religious and anthropological grounds; these voices found an echo during the Renaissance in the norms put into place to monitor access to the cadaver, to regulate the behavior prescribed for every aspect of the public anatomy lesson, and to endorse the prudent strategies adopted by anatomists. These strategies are concealed behind the rhetoric of the texts themselves and even behind the typographical makeup of some of them. Of course, I could not have identified the strategies without a preliminary analysis of the writings of the first critics of dissection. The extraordinary continuity between ancient and modern arguments and the qualms over practice is astonishing. The manner of their expression and the adoption of certain rituals in Renaissance anatomy suggests that they were generated by a common fear of contamination by proximity to the impure and malodorous bodies of the deceased in funerary practice. This contact affected the physician and the young student alike, and they managed to overcome their revulsion toward the cadaver only by an individual accommodation to it derived from the necessity of fulfilling the objectives of the profession. This attitude also affected society as a whole, which continued to hold the opening up of the body as a dishonorable and sacrilegious event. It is important to keep in mind these seemingly marginal aspects of the argument, because they effectively obstructed the possibility of conducting dissections and the means of acquiring knowledge through doing so.

In conclusion, it would seem to me that only by revealing the hidden, and yet significant, relationship between empirical science and human conscience, only by considering the problem of anatomy as it crosses the boundaries between epistemology and anthropology over a long period, can one come up with a full explanation of the problems this book seeks to elucidate. Just as the opponents of dissection intimated, a discussion of anatomical practice should not be separated from an analysis of broader cultural attitudes toward the body and toward death.

This is, after all, what concerns us most.

CHAPTER ONE

Representations: An Iconographic
Investigation of the Dissection Scene

The title page of a book is one of the few typological elements introduced by printing that is without precedent in the manuscript tradition.[1] It provides a new place, outside the text, in which the reader is given general information on the content of a book, on its author and publisher. Beginning in the sixteenth century this data was displayed in a rather rigid and codified form throughout Europe; at the same time it was accompanied by decorative elements or actual illustrations that added complementary facts concerning the content of the book as well as the activity of the author both to the title page and to the text. For the first part of the century these illustrative elements consisted of woodcuts, but they were subsequently more often followed by copper engravings, which peremptorily introduce the subject of the book, while remaining marginal to it.

Parallel to the establishment of anatomy as an academic subject, the anatomical lesson was frequently represented in the title pages of some medical and, more frequently, anatomical works published during these years. The illustrated title page, in effect, offers the most explicit and homogeneous description of the ceremony of the anatomy lesson shown at its most crucial and solemn moment, that of human dissection. The details contained in this page serve to interpret the actual procedure of the practice. By means of a figurative language made up of both explicit and of more hermetic elements, the title page reveals some of the action and some of the relationships that come into being while the ritual of the anatomy lesson is played out. The text itself never succeeds in describing

---

1. See, especially for the Italian books, F. Barberi, *Il frontespizio nel libro italiano del Quattrocento e del Cinquecento* (Milan, 1969).

the work of dissection and the morphology of the lesson itself as vividly. Sixteenth-century anatomical texts provide only rare rhapsodic details, or they leave unstated (as something that would be obvious to the supposed reader) any comprehensive description of the attitude of the physician toward the cadaver, the relationship between theory and practice in anatomical study, or the changing roles assigned to the public viewing of the dissections or the function of the instruments used in them. The visual representation of the anatomy lesson provides an initial insight into the secret of the cultural and technical attitudes implicit in the differentiated uses of dissection.

THE QUODLIBETARIAN MODEL:
THE TITLE PAGES OF MONDINO DEI LIUZZI'S "ANATOMIA"

One of the earliest printed representations of the anatomy lesson is contained in John of Ketham's *Fasciculus Medicinae,* a collection of medical texts issued for the first time in Latin in 1491 in Venice by Giovanni and Gregorio de Gregoriis, who also published an Italian translation by Sebastiano Manilio two years later.[2] The book was highly successful and it circulated widely, serving as one of the founding texts of academic medical and anatomical science at least until the mid-sixteenth century. It was a manual intended for medical students and barbers as well as for surgeons and physicians. It contained a number of fundamental late-medieval medical writings of a highly technical nature, including everything from Pietro da Montagnana's method "of judging urine by its color" to the technique for bloodletting; from the figure of zodiacal man with the correspondences between the planets and the parts of the body to advice on the plague by Piero da Tusignano; and, finally, the anatomy treatise by Mondino dei Liuzzi.[3] This text had been written in Latin in 1316 and

2. John of Ketham, *Fasciculus medicinae* (Venice: G. & G. de Gregoriis, 1491) (IGI, n. 5296a). The 1493 Italian translation is in IGI, n. 5300.

3. In the title page of the 1493 Venetian edition one reads: "Queste sono le cose contenute in questo dignissimo fasciculo di medicina volgare: el quale si contiene le sotto scripte cosse per sanità del corpo humano:

(1) El modo de judicar la urina per li soi colori de tutte le infirmità del corpo humano scripto in figura.

(2) El modo de trazer el sangue et sotto a che pianeto.

(3) La figura de l'huomo come le sottoposta ali pianeti.

(4) La figura della matrice tracta dal naturale.

(5) El consiglio per la peste del Maestro Piero di Tusignano.

(6) La anathomia de Maestro Mondino dechiarata de membro in membro.

(7) Virtú dalcune herbe secondo Plinio e Alberto Magno: et molti altri che hanno scritto."

printed for the first time in 1474.[4] In the 1493 Venetian edition of the *Fasciculus* the opening page of the treatise shows a dissection scene. We shall return later to the detail of the text and its sources; it is necessary first, however, to provide a proper context for the image. Mondino's is an anatomy manual in which the individual parts are presented in the very same order to be followed by the anatomist in dissecting the cadaver, beginning with the abdomen, which would be most susceptible to putrefaction, and finishing with the limbs. It was the first *mise-en-texte* of anatomy as it should be conducted. The text, which stresses that Mondino had conducted firsthand observations on the human body, is in stark contrast to the heavily Galenic character of the *Anatomia,* which is wholly inspired by the *De juvamentis membrorum,* a sort of compendium of Galen's anatomy that began to circulate in the Christian West at the end of the twelfth century.[5] Although Mondino had performed dissections with his own hands, in the *Anatomia* he fell prey to the same errors as did the great physician of Pergamum.

We know of seven different pictorial representations of Mondino's anatomy lesson that have appeared in the many reprintings of the *Fasciculus* and of the *Anatomia.* The first image chronologically is the one contained in the first edition of the Italian translation by Sebastiano Manilio of John of Ketham's *Fasciculus Medicinae* appearing in 1493 (fig.

4. Mondino dei Liuzzi, *Anatomia* (Padua: Petrus Maufer, c. 1474) (IGI, n. 5910).

5. This is a Latin version of an Arabic translation, abbreviated and simplified, of Galen's *De usu partium,* which differs considerably from the original. The *De usu partium* had already been translated into Syriac in the sixth century by Sergius of Resaina, but in the ninth century, Hunain, who was dissatisfied with this first version, made one of his own. The latter also completed and corrected a deficient Arabic translation that had been produced by his nephew Hubaish. It was probably this incomplete and abbreviated version of the *De usu partium* that was subsequently translated into Latin in the twelfth century with the title *De juvamentis membrorum.* This brief compendium summarizes, in ten books, twelve of the seventeen books that compose the *De usu partium.* Even as late as the high Renaissance it enjoyed much greater success than did the complete work. Despite the fact that Niccolò da Reggio had already translated the entire *De usu partium* in the first half of the fourteenth century, in the sixteenth century the *De juvamentis* appeared in certain editions of Galen's complete works: (Venice: P. Pincius, 1490); (Pavia: J. de Burgofranco, 1515–16); (Venice: L. A. Giunta, 1528) (cf. R. J. Durling, "A Chronological Census of Renaissance Editions and Translations of Galen," *Journal of the Warburg and Courtauld Institutes* 24 [1961]: 230–305). On the numerous manuscripts in which the *De juvamentis* appears, cf. D. Campbell, *Arabian Medicine and its Influence on the Middle Ages* (London, 1926), 2:43. The attribution of the *De juvamentis* to Taddeo Alderotti suggested by C. Singer is clearly incorrect: "The Confluence of Humanism, Anatomy and Art," in *Fritz Saxl 1890–1948: A Volume of Memorial Essays from his Friends in England,* ed. D. J. Gordon (London, 1957), p. 264. On these issues, cf. R. French, "*De juvamentis membrorum* and the Reception of Galenic Physiological Anatomy," *Isis* 70 (1979): 96–109.

1).[6] The illustration depicts an anatomy lesson as one might have observed it in those years at the University of Padua. In the foreground the body of a youth about to be dissected lies on a plank supported by two sawhorses. At the foot of the table is an object always shown in the iconography of the lesson: a basket for the detritus of the operation, which will be buried later with the rest of the body. At the center, knife in hand, stands the *sector,* the only figure not in academic gown, the one who will actually perform the dissection. He is leaning over the body and is about to make the first incision on the cadaver.[7] Standing at the extreme right, holding a pointer,[8] is the *demonstrator* or *ostensor,* who is indicating to the *sector* where and how to cut on the basis of what is being said or read by the person seated at the pulpit, the *lector.* The six other bystanders are students or instructors, as can be deduced from their garments. They stand around more or less distractedly, without entering into any apparent direct relationship to the cadaver.

This model of an anatomy lesson, which has been defined as quodlibetarian, predisposes that—in a hall that will shortly become a true and proper anatomy theater—the *lector* reads or recites from memory passages taken from the classical anatomical texts: the first *fen* of Avicenna's *Canone,* Galen's *De usu partium corporis humani* (both works had at this point replaced, at least in part, the *De juvamentis membrorum*), and Mondino's own *Anatomia,* which during the sixteenth century remained the most widely used university text for the study of anatomy in a strictly Galenic form. The *ostensor,* frequently translating the text from Latin into the vernacular, indicates to the *sector* the parts of the body that are to be dissected and displayed. The latter was generally a surgeon or, more frequently, a barber who certainly knew little or no Latin. The other six persons appearing in the illustration in academic gowns, who seem little interested in what is taking place before their eyes, are deep in conversation. They will be the ones to put into play the last phase of the *quodlibet:*

6. The same woodcut was used for the republication in Venice by G. & G. de Gregoriis, 1508. The attribution is doubtful. Singer ascribes it to a no better identified "Master of the Dolphins" in "The Confluence," p. 262. A. Hyatt Mayor (*Artists and Anatomists* [New York, 1984], p. 61) assigns it instead to Gentile Bellini, an attribution repeated in D. R. Karp, M. Cazort, et al., eds., *Ars Medica: Art, Medicine and the Human Condition* (Philadelphia, 1985).

7. It should be noted that the first incision, in accordance with Mondino's text, went from the point of the sternum to the pubes for the purpose of proceeding first of all to the dissection of the abdomen. This was the practice followed in all the dissections of the time.

8. This was the *radius,* an object that usually appeared among the iconographic attributes of the astronomer.

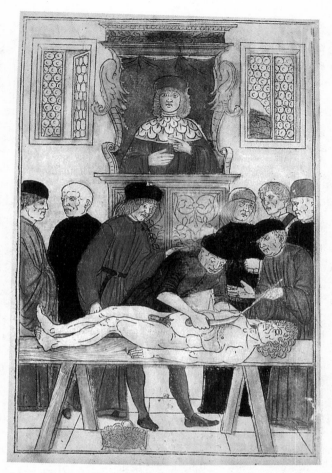

Figure 1   John of Ketham, *Fasciculus Medicinae* (Venice: Gregorio de Gregoriis, 1493).
John of Ketham's *Fasciculus medicinae,* first printed in Venice in 1491, is a collection of
texts and images intended to serve for both medical practice and for instruction. Mondino
dei Liuzzi's *Anatomia,* written in 1316, is included in this work and is introduced by a title
page, which contains a representation of the anatomy lesson. The *lector,* the person who
reads or recites the anatomical text, usually one inspired by Galen, is garbed in formal
vestments and is seated on the *cathedra,* which dominates the scene and is situated apart.
Eight figures, in addition to the cadaver, are portrayed in the lower part of the illustration.
A person standing to the right indicates with a pointer where and when to cut into the
sections of the body in accordance with what is being recited or read by the *lector.* He is
the *demonstrator* or *ostensor,* who plays a mediating role between the *lector* and the *sector*
(the latter physically performs the dissection of the cadaver), between word and deed. The
anatomy lesson as it is represented in this image corresponds to the normative descriptions
still being enunciated in the seventeenth century, for example, in the statutes of the University
of Padua (1607).

the *disputatio,* the discussion that will follow the reading of the text and the practical demonstration.[9] The *lector,* according to what had been decreed by the statutes of the University of Padua,[10] was normally selected from the ranks of the professors *straordinarii,* that is, from the youngest among the faculty, in order "to recite and read from Mondino's text of anatomy." As Jerome Bylebyl has suggested, according to the Paduan model it is the *demonstrator* and not the *lector* who is the physician presiding over and directing the anatomy lesson. This role was in fact performed by one of the four full professors of medicine (whether theoretical or practical was of no importance). It was his duty to explain the passage that had just been read and to "demonstrate and verify it faithfully on the cadaver."[11]

In spite of the information provided by the 1465 Paduan statutes, it seems to me that the title page offers contrary evidence on the identification of the figure guiding the anatomy lesson. In the image the emphasis falls with particular force on the central role played by the *lector* as the orchestrator of the entire performance. I have deduced this from his more solemn academic garments, as well as from the preeminence of his position in the organization of the image. This hypothesis is supported by the evidence in the statutes of the College of Physicians in Rome, where the role of the reader surpasses that of "the two censors" who have to be elected by the members of the college to fulfill their positions as *sector* and *ostensor.*

Successive editions of Mondino's work contain illustrated title pages taken from woodcuts. These woodcuts frequently differ from the one used in the 1493 version. The basic framework of the image is generally the same, given that successive cuts were all copied, though sometimes reinterpreted, from the preceding edition. The differences between them can be discerned in certain significant details precisely because we are dealing with a reworking of the original design, and they provide information about successive readings of an iconographic text.

Of the revisions of the 1493 woodcut, the one that has remained most faithful to the first printing is without doubt the title page of the Venetian

9. The "quodlibetarian model" is the definition used by W. Heckscher, *Rembrandt's Anatomy of Dr. Nicolaas Tulp: An Iconological Study* (New York, 1958). He describes the *quodlibet* as: "the sophisticated public disputes that, from the thirteenth century onward, had become, as it were, the show windows through which the non academic outsider could observe and enjoy the goings-on of the universities" (pp. 45–46).

10. *Stat. Gymn. Pat.,* L. II, chapter 28. See chapter 2 for an analysis of these 1465 statutes.

11. Ibid. Cf. J. J. Bylebyl, "The School of Padua: Humanistic Medicine in the Sixteenth Century," in *Health, Medicine and Mortality,* ed. C. Webster (Cambridge and New York, 1979), p. 353.

1522 edition of the *Fasciculus Medicinae* (fig. 2).[12] It contains only one or two significant changes. First among them are two alterations in the figure of the *lector:* here he is represented no longer as a youth, but as a mature man with an expression of greater authority compared to the 1493 title page. This would seem, at least in part, to contradict Bylebyl's hypothesis, which is supported by the model of the lesson derived from the Paduan statutes, since the *lector* now is not necessarily the youngest of the professors, recruited from the part-time faculty of the university. Moreover, the book appears on the lectern: in fact, of all the known versions of this image, only the first, the 1493 edition, shows the reader reciting. The presence of the book, juxtaposed with that of the cadaver in the iconography of the dissection scenes, is constant and always highly significant in exemplifying the interconnection between the theory and practice of anatomy. Finally, another detail in this image needs to be stressed: two figures now appear among the bystanders, and they follow the anatomical demonstration with attention.

A grotesque reworking of the title page of Mondino's *Anatomia* appears in the 1514 Rostock edition.[13] However, the iconographic structure remains substantially unchanged in respect to the other two figures: the *lector* is on his *cathedra* and has a large volume before him; the *incisor* has cut the cadaver in a line running from the pubis to the sternum; the two figures, who with the *sector* occupy the space of the dissection area, direct their gaze and entire attention in the direction of the teacher, who is gesturing to the public.

In the edition of the *Fascicolo di Medicina* published in Milan in 1509 (fig. 3) we find a woodcut that is a reduced replication of the 1493 version.[14] In it the *lector* is holding a pen in his right hand, and as many as

12. Printed on December 31 by Cesare Arrivabene. Another edition, resembling the preceding one, dated January 7, 1522, is recorded by M. Sander, *Le livre à figures italien (depuis 1467 jusqu'à 1530)*, 6 vols. (Milan, 1942–43). This engraving had been used previously by de Gregoriis for the following editions of the *Fasciculus:* 1495 (IGI, n. 5297); February 17, 1500 (IGI, n. 5298); March 28, 1500 (IGI, n. 5299); and 1513.

13. Mondino dei Liuzzi, *De omnibus corporis humani membris interioribus anatomia cum figuris faberrimis non solum medicis sed philosophantibus etiam omnibus utilissima* (Rostock: Nikolaus Marshalk, 1514). I have not succeeded in obtaining a reproduction of this image.

14. John of Ketham, *Fasciculus Medicinae* (Milan: Giovanni da Castiglione, 1509). This is a forgery of the Venetian 1493 edition. The illustrations are all crude, reduced copies from this edition, which was republished the following year, again in Milan, by Io. Aug. Scinzenzeler. M. Sander (*Le livre*) cites an extremely rare 1516 edition of the *Fasciculus* by the same Milanese printer, with illustrations attributed to Andrea Mantegna. Sander's source is the *Catalogue de la bibliothèque de feu M. le Comte Jacques Manzoni*, 4 vols. (Città di Castello, 1892–94).

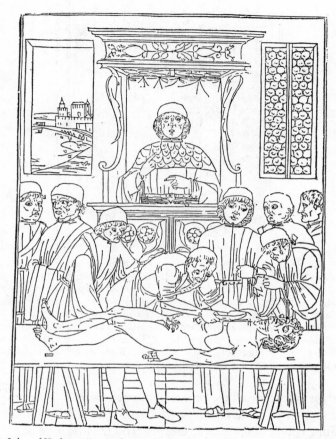

Figure 2   John of Ketham, *Fasciculus Medicinae* (Venice: Cesare Arrivabene, 1522; Bibliothèque Nationale, Paris). The scene that is used as a model for the anatomy lesson in the
Venetian edition of the *Fasciculus Medicinae* had previously appeared in the 1493 imprint.
The composition of the image is the same. Only a few details are different: the basket at the
feet of the dissecting table is gone; without the use of a pointer, the *demonstrator* provides
the *sector* with instructions on how to open up the cadaver; on the *lector's* dais a book
is now visible. The operations being performed on the cadaver follow the reading of the
authoritative text and serve to demonstrate what is described in it. The dissection—in the
context of the public anatomy lesson and the regulations of the university statutes—is
shown exclusively as a visual aid to the words pronounced ex cathedra.

four others, beside the *sector* and the *demonstrator,* appear to be following the activity around the cadaver on the dissecting table attentively, commenting on what they are observing. Meanwhile two figures on the right
are conversing intently, and seem completely detached from the central
scene. In the 1519 Geneva edition of the *Anatomia Mundini,* however,
everyone gazes at the cadaver, and the figure of the *lector* is disproportion-

Figure 3   John of Ketham, *Fasciculus Medicinae* (Milan: Giovanni de Castellione, 1509). In the Milan 1509 edition of the *Fasciculus Medicinae* the depiction of the anatomy lesson resembles the one in the 1493 Venetian imprint. The *lector* is represented with a pen in his right hand, as if to underline further his intellectual role and his removal from the practical act of dissection. In this case, four figures, in addition to the *sector* and to the *demonstrator,* seem intent on pursuing the progress of the dissection. One of these personages, who is present in many versions of this image, including the 1493 edition, places his hand on the *sector's* back, as if to encourage him. Dissecting a cadaver was a problematic operation from many points of view: the anthropological, the religious, the moral, and the emotional.

ately larger than that of the others and, once again, has an open book before him (fig. 4).[15] The volume is the size of a large folio and becomes a dominating coprotagonist of the engraving on the title page of the French translation of the *Anatomia* published in Paris in 1532 (fig. 5).[16] The *lector* is in a different position as compared to the one assumed in the cases described previously: the lectern has become a sort of pulpit, and its distance from the body and the scene of dissection has grown enormously. The *lector,* upon completing the reading of the text, points with his index finger and turns his gaze on the cadaver, on which the attention of all the physicians and students converges. Here the text will find its confirmation.

The final version of a title page to Mondino's *Anatomia* that should be considered is that of the Leipzig edition prepared by Martin Pollich, c. 1495 (fig. 6).[17] Although it precedes the others chronologically, it is

15. Mondino dei Liuzzi, *Anatomia* (Geneva, 1519).

16. Mondino dei Liuzzi, *Anatomia* (Paris: A. Lotrian et D. Janot, 1532).

17. Mondino dei Liuzzi, *Anathomia Mundini. Emendavit Martinus Mellerstat* (Leipzig: Martin Landsberg, c. 1493) (IGI, n. 5914a), where Martinus Mellerstat stands for Martin Pollich of Mellerstadt in Franconia. Martin Pollich is identified in R. Herrlinger, *A History*

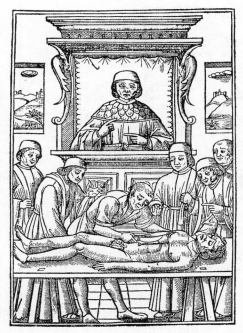

Figure 4   Mondino dei Liuzzi, *Anatomia* (Geneva, 1519). Mondino dei Liuzzi's anatomy, in addition to appearing in all the many editions of John of Ketham's *Fasciculus medicinae,* was also published autonomously. It was printed for the first time in Padua by Pierre Maufer, probably in 1474. After that, many Latin and vernacular translations appeared throughout Europe during the fifteenth and sixteenth centuries. In the Geneva 1519 version the title page reproduces the model of the anatomy lesson represented in the Venetian editions of the *Fasciculus medicinae.* Here, the figure of the *lector* and the space occupied by the *cathedra* are disproportionately larger in the whole design of the illustration, probably to emphasize the preeminence of bookish knowledge—founded on ancient authorities and dispensed orally in the lesson—and to distinguish it from the more strictly manual aspects of the anatomy. Just as in many other versions of this scene, the gaze of the *lector* is clearly directed to a point outside the image, beyond the cadaver and the other figures represented in the lower part of the title page. On the other hand, those figures in the Geneva edition are all attentively observing the dissection.

considered last since it represents a rather radical reworking of the original image. To begin with, it is an unspecific but condensed interpretation of the contemporary anatomy lesson as it was then and as it would be shown later. The lettered title is set in the woodcut title page, somewhat in the manner of those first incunables that were modeled on illuminated manuscripts. The entire illustration, moreover, in terms of both the qual-

of Medical Illustration from Antiquity to 1600 A.D. (Nijkerk, 1958), p. 62, who was the first to publish this title page.

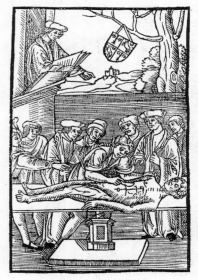

Figure 5 Mondino dei Liuzzi, *Anatomie* (Paris: A. Lotrian and D. Janot, 1532). Some late medieval miniatures reveal that dissection and autopsies occasionally took place out-doors. This title page of a French edition of Mondino's *Anatomia* (Paris: A. Lotrian and D. Janot, 1532) seems to be referring to this icono-graphic tradition. This time the *lector* is defi-nitely placed on a pulpit rather than on a *cathe-dra*, and the book is a large (and awkward) folio: together, they dominate the scene of dis-section, which takes place in the lower part of the engraving. The bystanders, along with the *lector* himself, are all gazing toward the ca-daver, the place where bookish knowledge finds its empirical verification.

ity of the woodcut and of the organization of the image, is of rather crude workmanship and is still notably medieval in aspect. Only three figures are represented in the rustic landscape: the *lector,* the cadaver, and the *sector.* The *lector,* seated on a lectern, dominates the scene, the anatomy text in his left hand, while he points to the cadaver stretched out on the table in the foreground with the index finger of his right hand. Here the *lector* has also taken on the part of *demonstrator,* since his role is that of linking the spoken word proffered from the lectern and the anatomical demonstration in that limited sense that was described before. The *sector* looks toward the professor and carries out his orders; his inferior position in terms of age, dignity, and function is quite evident in this image. He is represented inserting his hands into the cadaver's abdomen after having made the original incision, following the teacher's instructions.

At this point a number of observations should be made regarding the title pages as a whole, the interpretation of the dissection, and the anat-omy lesson that they propose. In spite of the type of anatomical demon-stration established by the Paduan statutes of 1465 (namely a lesson directed by a full professor in the role of *demonstrator,* assisted by a younger instructor charged with the reading of the anatomical text), it would seem that the 1493 title page, as well as all those following, deci-sively emphasize the preeminent function of the *lector.* His position in the iconographic organization of all the above images, in a space defined by the professorial pulpit and clearly distinguished from the opposite space for dissection, makes evident the scission between the theoretical activity

Figure 6   Mondino dei Liuzzi, *Anathomia . . . emendata per doctorem melestat.* (Leipzig, 1495; Biblioteca degli Istituti Ortopedici Rizzoli, Bologna). The author of this title page in an "emended" German edition of Mondino's *Anatomia* furnishes a fabricated version, so to speak, of the anatomy lesson. The scene unfolds in a rural setting, and only the three essential figures are represented: the *lector* who, above in the *cathedra* and holding a book in his hand, points out to the *sector* the procedures to be performed on the outstretched body lying on the dissecting table. The hierarchical relationship between the two personages is strongly underlined by the composition of the image, as well as by the fact that the *sector* is portrayed as a youth who carefully attempts to fulfill the instructions delivered to him from the *cathedra*.

of the physician-anatomist and the practical example directed toward an empirical examination of the cadaver. This dichotomy between theory and practice can be seen in a schematic and simplified though relevant way on the title page of the Leipzig 1495 edition.

This is indeed the essential subject for reflection offered by the series of title pages of Mondino's *Anatomia*. Mondino's text is a sort of compendium to Galenic anatomy, but it also offers important technical information on the dissection procedure, occasionally emphasizing its manual aspects. The text alludes to two dissections performed on the human body by the author himself in January and March 1315, and also refers to scores of others executed on animal bodies. It appears that, thanks to Mondino, an old inconsistency that went back at least to the extinction of the school of Alexandria founded by Herophilus in the third century B.C. was laid to rest. This inconsistency took the form of a contradiction between an anatomy based on texts and another anatomy which, by force of statute, was obliged to verify the information imparted by the texts through a direct observation of the dissected cadaver. These expectations, though, are disregarded by the *Anatomia*. In spite of his observations on the human body, and in spite of the dissection, Mondino's text slavishly retraces the footsteps of Galen and Avicenna, assuming all their imperfections and errors, including the crudest (such as, for example, the description of a five-lobed liver and a tripartite sternum), which a careful investigation based on the cadaver could easily have dissipated. The practice of dissection, consequently, did not necessarily expose ancient errors, and this was due in large part to the methods according to which it was carried out. The title pages that adorn Mondino's editions furnish a visual description of the use of dissection and its purposes in academic instruction. The anatomy lesson, in the light of this model, turns out to be little more than a ritual to celebrate the ancient classical authorities on the subject through a reading of their texts and the *disputatio,* which, in addition, never truly questioned the authority of the writers and the content of their works. Dissection itself, inserted in the *quodlibetum,* became an integral part of academic ritual, deprived of all its innovative potential. The title pages, especially the early versions, demonstrate with sufficient clarity the limited interest on the part of the public of students and physicians and the *lector* himself in the anatomical demonstration, which is evidence of the scientific deficiency of the procedure. The dissections depicted in these representations did not imply exploring the interior of the body and distinguishing individual parts and their interlocking function through direct observation. The purpose of dissection gathered from these title pages is strictly didactic: to show what the text described and to offer visual support for the words of the reader. Even more than this the anatomy lesson assumes the features of one of the many *quodlibeta* celebrated by the other disciplines: instead of offering the opportunity for scientific enrichment through the publicizing of new discoveries and teachings, performances were destined to affirm (or reaffirm) the authority of the classical sources of knowledge through public discussion, and at the same time, the authority of the academy that administered and dispensed them. Indeed, the series of representations of the anatomy lesson taken from the title pages of Mondino's *Anatomia* appear precisely as celebrations of the classical authors (Galen, Avicenna, and, in this case, Mondino himself) and of the *lector* as the academic conduit for the diffusion of knowledge. The person of the physician-anatomist therefore is substantially demoted. He becomes merely a student of texts, and the practical aspect of teaching, and even more of anatomical research, remains totally secondary to his purpose. At the same time, the dissection of the cadaver by the *sector* becomes a ritualized exercise void of any significant investigative aim.

The Persistence of a Model: Berengario da Carpi

The many reeditions of Mondino's *Anatomia* and the fact that it was adopted as a university anatomy manual testify to its success. In the course of the sixteenth century numerous works of anatomy offered commentaries to it or were inspired by it. Iacopo Berengario of Carpi, lecturer in medicine and surgery in Bologna, was without doubt the best known and in certain respects the most important commentator on Mondino. In 1521 he published his *Commentaria cum amplissimis additionibus super anatomiam Mundini una cum textu ejusdem in pristinum et verum nitorem redacto*.[18] In this large work totaling 528 leaves Mondino's much briefer *Anatomia* is subjected to a learned and minute scrutiny through a comparison with other writers on anatomy, ancient and modern. The large number of textual references contained in the *Commentaria* permits the reconstruction of a complete bibliography of anatomical works in circulation up to the date of Berengario's own text. However, the verification of Mondino's text and the commentary follow an observation of the dissected cadavers (to which the author frequently alludes), and he insists on the priority of sensory perception over traditional authority, Galen's included. Berengario writes: "and so let them be cautious in considering books of anatomy and not believe authorities, but discern the matter as we do."[19] Alongside this display of a systematic critique based on authoritative anatomical sources, and taking his cue from the manual of Mondino dei Liuzzi—already a profound innovation in the scholarship devoted to the discipline—Berengario's work contains twenty-one images, of a generally rather crude and imprecise appearance, which illustrate the text.

From both the epistemological and cultural as well as the morphological point of view, the *Commentaria* differs profoundly from the work that inspired it: the evidence of a repeated and constant return to the dissection of the human body, the severe and punctilious critique of classical authorities, and the use of illustrations are all elements that signal a change of attitude on the part of the physician with regard to anatomical knowledge and the ways it is to be pursued and transmitted. In spite of this, the title page of the 1521 Bologna edition of the *Commentaria* (fig. 7) reproduces the scene of the anatomy lesson and of the dissection in a manner that substantially resembles that of the illustrations of Mondino's anatomical work. The title is printed in red letters between two columns and is dominated, in the center, by the coat of arms of the Medici family

18. I. Berengario da Carpi, *Commentaria cum amplissimis additionibus super anatomiam Mundini una cum textu ejusdem in pristinum et verum nitorem redacto* (Bologna: Hieronymum de Benedictis, 1521).

19. I. Berengario da Carpi, *Commentaria cum amplissimis additionibus super anatomiam Mundini una cum textu ejusdem in pristinum et verum nitorem redacto* (Bologna: Hieronymum de Benedictis, 1521), fol. CLIIIv. Cf. L. R. Lind, *Studies in Pre-Vesalian Anatomy: Biographies, Translations, Documents* (Philadelphia, 1975), p. 10.

(the work is dedicated to Cardinal Giulio de' Medici) and, at the upper edges, by the name of the reigning pope, Leo X, a member of the family. The initials of the printer, Hieronymus de Benedictis, are inscribed at the base of the columns. Below, the woodcut reproduces an anatomy lesson. The *lector* is no longer seated on a rostrum, and it would appear that the distance separating him from the dissecting table has been reduced in comparison to earlier title pages. Nevertheless his role remains strictly that of the verbal expounder of the anatomical lesson, while the *sector* executes the incisions on the cadaver and, knife in hand, separates the individual parts to be displayed. The *ostensor,* on the other hand, is no longer in the picture.[20]

In 1522 Berengario published another anatomical work entitled *Isago-gae breves perlucidae ac uberrimae in anatomiam humani corporis . . . ad suorum scholasticorum preces in lucem datae.*[21] The book consists of a brief text of seventy-two leaves in which the findings of the *Commentaria* are condensed. The illustrative apparatus has remained substantially the same as in the previous work.[22] This treatise, both bibliographically and from the point of view of content, as the title itself states, is a manual for the medical student.[23] The *Isagogae* had been tacitly conceived by Berengario as a replacement for Mondino's treatise, which he considered largely superseded.[24] These features permitted Berengario's

20. The same woodcut was used many years later as the title page for Ulrich von Hutten's *De guaiaci medicina et morbo gallico liber unus* (Venice: Zilettus, 1567). It is quite curious that a scene of dissection should serve as an iconographic introduction to a book on syphilis. Even stranger is the fact that it kept both the initials of the publisher of the *Commentaria* inscribed on the columns, as well as the name of Leo X and the coat of arms of the Medici family. Ziletti may have had nothing better available.

21. I. Berengario da Carpi, *Isagogae breves perlucidae ac uberrimae in anatomiam humani corporis . . . ad suorum scholasticorum preces in lucem datae* (Bologna: Benedictum Hectoris, 1522).

22. In the *Isagogae* the figures numbered 14, 16, and 17 are missing. Figure 13 of the *Commentaria* was changed, and a previously unpublished woodcut appears at fol. 25r. For a useful synthesis of the bibliographical problems concerning the illustrations in the works of Berengario, cf. L. Choulant, *History and Bibliography of Anatomic Illustration,* ed. and trans. M. Frank (New York, 1945), pp. 136–42; 1st ed. [German] Leipzig, 1852.)

23. The references to other contemporary authors have been eliminated and those to ancient authorities have been reduced to the essential. Berengario conceived of his *Isagogae* precisely as a compendium, an epitome of the *Commentaria,* and to this work he refers readers of the *Isagogae* who desired a more detailed and articulated discussion of the parts of the human body (cf. *Isagogae,* 1522 ed., fol. 2r).

24. Although Berengario considered Mondino's *Anatomia* an authoritative source for anyone who was embarking on the study of the human body, he nevertheless pointed out its limitations: "Veritas est quod Mundinus in aliquibus deviat a veritate et etiam in nonnullis mancus reperitur," quoted from I. Berengario da Carpi, *Commentaria,* fol. IIIv.

Figure 7   Iacopo Berengario da Carpi, *Commentaria . . . super Anatomiam Mundini. . . .*
(Bologna: Girolamo de Benedictis, 1521; Biblioteca Angelica, Rome). In 1521 Iacopo Berengario da Carpi, reader in medicine and surgery in Bologna, published there a long and punctilious comment on Mondino's *Anatomia*. Mondino's brief text is analyzed, discussed, and compared to other sources in more than 500 leaves. The work is dedicated to Giulio dei Medici: the coat of arms of the family and the name of Leo X (a Medici) appear on the trabeation of the architectonic frame to the title. In the anatomy lesson represented in the lower part of the image, the scene is, once again, based on the quodlibetarian model: a *lector* who orally describes the anatomical parts, a *sector* who demonstrates them, and a group of persons who will carry out the *disputatio*. This discussion is the final part of the *quodlibetum,* the academic ritual within which the dissection occurs, a ritual better suited for the celebration of the authority of the book than for a true discussion of what has been heard or seen.

text to circulate much more widely than the *Commentaria*,[25] although it never succeeded in displacing Mondino's *Anatomia* from its primary position as an anatomical manual. The *Isagogae* was in fact reprinted the following year by the same publisher, Benedictus Hectoris, and another edition appeared in Venice the same year. They were followed by three editions in Strasbourg (in 1528 together with the anatomy of Alessandro Benedetti, and again in 1530 and 1533), and a Venetian one printed in sixty-three leaves in 1535 by Bernardino de Vitali.[26]

For the Bolognese 1523 reprinting of the *Isagogae,* the publisher reused the woodcut from the title page of the 1521 *Commentaria,* introducing only the following variants: "Carpus" is inscribed under the figure of the *lector,* thereby identifying Berengario himself as the teacher; *Carpus* is also inscribed on the trabeation over the columns; and the initials of the author (I. Be.) replace those of the publisher at their base.

In the Venetian 1535 edition of the *Isagogae* printed by Luca Antonio Giunta the title page still includes a representation of the anatomy lesson (fig. 8). In this case, the representation consists of a reutilization of a woodcut that Giunta had used twelve years earlier for an edition of the works of Ugo Benzi.[27] It is a mere reworking of another print that a certain Lunardo had produced in 1522 for an edition by the same publisher of Galen's *Therapeutica* (fig. 9).[28] The author of this image, at least as far as the general organization of the dissection scene is conceived, must certainly have been influenced by the title page of the 1521 *Commentaria.*[29]

The scene offered by the title page of the 1535 edition of the *Isagogae* is much more complex in respect to previous exemplars, and the definition of roles is certainly more rigid, especially if measured against the content of the text and its technical assumptions:

(1) A figure holding a vessel is at the side of the *sector.* His posture and dress seem to indicate that he is assisting the dissector. His presence therefore further delineates the roles of the protagonists of the *quod-*

25. The only edition of this text is that of 1521.

26. L. Choulant, *History and Bibliography,* pp. 138–41.

27. U. Benzi (U. Senensis), *Opera* (Venice: L. A. Giunta, 1523). I am grateful to Monique Kornell for the reference. On this question, see G. Wolf-Heidegger and A. M. Cetto, *Die anatomische Sektion in bildlicher Darstellung* (Basel and New York, 1967), n. 123.

28. Galen, *Therapeutica* (Venice: L. A. Giunta, 1522). The title page reads "Lunardus fecit."

29. An exact copy (but with a different woodcut) of Lunardus's title page by an anonymous engraver was printed in Avicenna's *Canon opera. Edidit Andreas Alpagus* (Venice: L. A. Giunta, 1527) (fig. 10).

libetum. The functions and hierarchical relations of the figures represented are consequently defined with greater emphasis than they are
in the title pages of Mondino's *Anatomia* or in those of the prior
editions of Berengario's works.

(2) The six listeners, all wearing togas, are seated on benches placed on
a platform at the same level as the teacher/reader's own seat. The platform defines the space of the lecture, a space that has priority and is
separate from the one in which the anatomical table (the space of the
procedure) is placed. In addition to the definition of the spaces and
of the specific functions that cover the practice of dissection, we can
observe the demonstration, reading, and discussion, and even the precise elaboration of the procedures that regulate "the seating arrangement following an order of dignity."[30]

This is the last in the series of images that represent dissection according to the quodlibetarian model, and these two elements suggest, to
me at least, an increasingly inflexible representation of the anatomy lesson—a representation more and more exclusively set in an iconographic
tradition that dictates the organization of the image and separates it from
its actual context. If, in fact, there is a disparity between the text of Mondino's *Anatomia* and the series of title pages at the beginning of the volume in its many editions, there is an even more pronounced discrepancy
between the works of Berengario da Carpi and the representations of the
anatomy lesson that they contain. The clear-cut distinction between the
roles, the separation of reading space and dissection space, the conflicting
relationship between theory and practice—all these elements define representation according to the quodlibetarian model, and they clash forcefully with the procedural statements found in the text. Berengario's
works, in fact, appeal constantly to the use of dissection and to empirical
verification. The model for teaching and learning anatomy that he proposes therefore distinguishes itself radically from that suggested by his
immediate predecessors and from that represented in the illustrations that
paradoxically introduce the various editions of the *Commentaria* and of
the *Isagogae*. Textual and academic authority is modified by the practice
of a direct observation of cadavers. The good anatomist, wrote Berengario, "does not believe anything in his discipline simply because of the
spoken or written word: what is required here is *sight and touch.*"[31]

30. A. Benedetti, *Anatomice sive historia corporis humani. Ejusdem collectiones medicinales seu aphorismi* (Venice: Bernardino Guerraldo Vercellensis, 1502), fol. aIII. On this
point, see W. Heckscher, *Rembrandt's Anatomy,* p. 46.
31. I. Berengario da Carpi, *Commentaria,* fol. VIv.

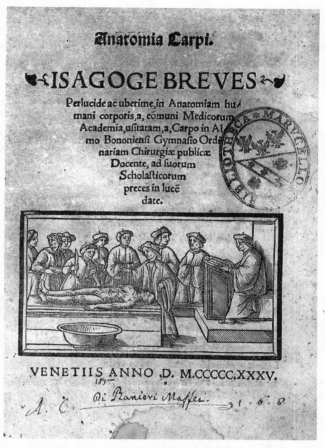

Figures 8–10   Figure 8: Iacopo Berengario da Carpi, *Isagogae breves* . . . (Venice: Lucantonio Giunta, 1535; Biblioteca Marucelliana, Florence). Figure 9: Galen, *Therapeutica*
(Venice: Lucantonio Giunta, 1522). Figure 10: Avicenna, *Canon opera* (Venice: Lucantonio
Giunta, 1527). In an attempt to face up to the exigencies of teaching and to students' needs,
but also with the expectation—promptly shattered—of being able to replace Mondino's
text as a manual of anatomical instruction, Berengario published in 1522 an illustrated
synthesis of his prolix commentary to Mondino: the *Isagogae Breves*. The dissection scene
appearing on the title page of the Venetian 1535 edition published by Lucantonio Giunta
demonstrates another way in which the separation between the space of the word and the
space of the manual labor is expressed graphically in the teaching of anatomy: it is marked
by the platform on which the *lector* and the other robed figures are placed, seated in precise
order ("pro dignitate"). This image adopts the iconographic model of the anatomy lesson
produced by a certain Lunardus for an edition of Galen's *Therapeutica,* also used by the
Giunta in 1522 (fig. 9) and again in the 1523 edition of Ugo Benzi's *Opera* and in 1527 for
Avicenna's *Canon* (fig. 10).

A Transitional Iconography?

In terms of iconography the continuity and viability of the quodlibetarian model were impaired by the title page contained in the Latin edition of Galen's *De anatomicis administrationibus,* translated by Johann Winther von Andernach (Johannes Guinterius Andernachus), printed in Paris in February 1531 by Simon de Colines, and reissued only a few months later (May?) in Basel (fig. 11).[32] The same publisher had already used this title page during the previous year in an edition of another work by Galen, the *Methodus medendi,* in the translation by Thomas Linacre.[33] The first Latin translation by Demetrius Chalcondylas of the *De anatomicis administrationibus* dates from the end of the fifteenth century and was printed with the title *De anatomicis aggressionibus* in 1529 in a collection

32. Galen, *De anatomicis administrationibus. De constitutione artis medicinae. De theriaca ad Pisonem. De pulsibus ad medicinae candidatos liber. Per Joan. Guinterium Andernacum Latinitate jam recens donata* (Basel: A. Cratander, 1531). All these texts are in Winther's translation from the Greek.

33. Galen, *Methodus medendi* (Paris: S. de Colines, 1530). For an overview of Colines imprints, see P. Renouard, *Bibliographie des éditions de Simon de Colines 1520–1540* (Paris, 1894). This title page has been attributed to Jollat, one of the authors of the figures contained in C. Estienne (Carolus Stephanus), *De dissectione partium corporis, una cum figuris et incisionum declarationibus a Stephano Riverio expositis* (Paris: apud Simonem Colinaeum, 1545). Cf. A. Bernard, *Geoffroy Tory, peintre et graveur, premier imprimeur royal* (Paris, 1865) and R. Herrlinger, *History of Medical Illustration* (London, 1970), p. 164.

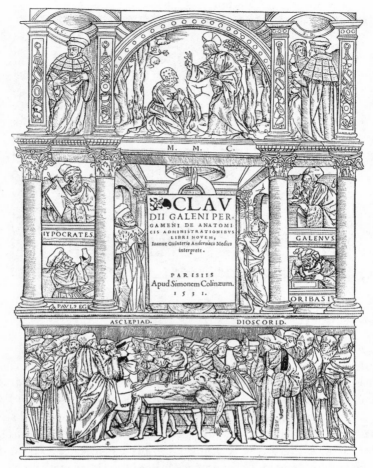

Figure 11   Galen, *De anatomicis administrationibus* (Paris: Simon de Colines, 1531). Johann Winther was one of Vesalius's teachers in the Paris medical faculty. In 1531 Winther published a translation of Galen's *De anatomicis administrationibus* that, in the Simon de Colines edition, opens with a richly illustrated title page. In the upper center there is a representation of the miracle of Christ's healing of the leper, and on the two sides, the image of the physician-saints Cosmas and Damian, one holding a book, the other a phial of urine, to signify the twofold connotation of medicine, theoretical and practical. The central part contains the portraits (obviously fanciful) of six founding fathers of medicine: Hippocrates, Galen, Paul of Aegina, Oribasius, Aesculapius, and Dioscorides. The composition of the dissection scene portrayed in the lower part of the title page is unusual. There is no longer a figure reciting; the separation between the space for theory and the space for practice has vanished; an agitated discussion appears to be in progress; two personages—probably students—are handling the anatomical parts: one of them with dramatic gestures turns to a person dressed in solemn academic garments standing at the head of the cadaver. The quiet composure of the anatomy lesson found in scenes on the quodlibetarian model has disappeared in this image.

of Galen's texts published in Bologna under the editorship of Berengario da Carpi.[34] Winther embarked on his new translation because of his deep dissatisfaction with the earlier version, which he judged to be obscure, crude, and often unintelligible, at least as far as some of the translations by Niccolò da Reggio were concerned.[35] Winther had studied and later taught Greek at Louvain. After he moved to Paris in 1527 he taught medicine there for a decade as the faithful transmitter of the most obstinate Galenic tradition.[36]

Nevertheless, the title page of his translation of the *De anatomicis administrationibus* marks a profoundly different and innovative development. The illustration can be divided into three parts, each of which possesses its own internal thematic consistency. At the top of the page on the right is St. Cosmas holding a book (signifying the study, the theory of medicine) and on the left St. Damian holds a phial (as a symbol of medical practice).[37] Between the two saints there is an image of Christ healing a leper.[38] In the central section of the title page are representations of the classic authorities of medicine: in the central panel Aesculapius and Dioscorides are shown as full figures with their gaze directed to a caption containing the bibliographical information on the text. Aesculapius is one of the founders of the healing arts, and Dioscorides is the author of the *De materia medica,* which remained the basic pharmaceutical text

34. Galen, *Libri anatomici* (Bologna: G. B. Phaelli, 1529). The collection contains: *De motu musculorum* (translated by Niccolò Leoniceno), *De anatomicis aggressionibus* (translated by Demetrius Chalcondyles), *De arteriarum et venarum dissectione* and *De nervorum dissectione* (translated by Andrea Fortolo), *De hirundinibus,* etc. (translated by Ferdinando Balamio Siculo). For editions of Galen in the Renaissance, see the valuable article by R. J. Durling, "A Chronological Census."

35. "Si exordia librorum excipias, reliqua tam obscure vertit, ut a nemine queat intelligi dum videlicet schemata graeca pugnante Latinorum idiomate reddere totidem verbis nititur, idque imperitus quemadmodum ... Nicolaus Calaber quibusdam operibus factitavit. Verum nos authoris mentem fideliter exprimere conati sumus, nunc ad verbum, nunc ad sensum, ne latinitati vim adferremus: atque haec omnia sermone simplici, et quantum licuit puro" (Galen, *De anatomicis,* 1531 ed., introduction). Niccolò da Reggio was one of the great translators of Galen: roughly between 1308 and 1345 he produced Latin versions directly from the Greek of some fifty texts.

36. His anatomy manual is explicit from its very title: J. Winther, *Institutionum Anatomicarum secundum Galeni sententiam ad candidatos Medicinae libri quatuor* (Basel: B. Lasium and T. Platter, 1536). The author clarifies in the preface: "Ego itaque, ut vere medicinam discentibus gratificam, quae hactenus ex Galeno in ipsis operibus vera esse comperi, breviter in quatuor libros contraxi, eo docendi ordine et ratione quam et Galeno probari video, et facillimam Medicinae candidatis fore speravi" (p. 9), and, a little beyond, "nihil hic scriptum est quod non Galenum redoleat, cuius doctrina ex professo imitor" (p. 11).

37. Saints Cosmas and Damian are recognizable through the context, by their material attributes, and by the initials S. C. and S. D. inscribed in the niches.

38. This episode is recounted in the New Testament (Matthew 8: 2 ff.; Mark 1: 40 ff.; and Luke 5: 12 ff).

throughout the sixteenth century. Between the columns are the portraits of Hippocrates, Galen, Paulus Aegineta, and Oribasius, all with a book before them, in the act of reading, studying, or teaching.[39] Finally, below, a heated discussion is taking place around the anatomical table.

This image, which is rather complex compared to the examples so far examined, synthesizes one way of understanding medicine. The physician saints Cosmas and Damian personify in the hagiographic tradition the Christian exercise of medical practice and offer protection to whoever approaches it in a spirit of devotion. The scene of the miraculous healing of the leper further reinforces the idea of medicine as a merciful practice. The figure of Christ healing through his own supernatural powers, in this context, can be understood as a metaphor for the physician's function. Christ is here represented, out of consideration for the goals that he sets for himself and the results that he achieves (certainly not by the tools he uses) in the guise of a physician. The ancient authorities are represented in the weight-bearing portion of the architectonic structure, as if to underline the indispensable role they play in medicine and to emphasize a major aspect of the physician's activity. The scene of the dissection and of the discussion involving teachers and pupils shown in the lower part of the title page completes this vision of medicine and represents its more manual aspects.

The anatomy lesson, however, is portrayed here in a rather unusual way, which needs some clarification. One is instantly struck by the absence of a boundary between the space where the dissection takes place and the academic space, that of oratory and bookish knowledge (to use the terminology employed in the description of the quodlibetarian model). In a crowded hall filled with some thirty or more people an animated discussion involving professors and students is underway. In the center foreground is the object of their reflections: a body with its abdomen gaping. The right hand of a youthful student (no longer a *sector*) is immersed in the viscera of the cadaver, and the student's left hand is raised as if to reinforce a statement made to a person in solemn professorial vestments. The latter seems to be urging propriety and caution on the young man. Also in the center, slightly to the left, another youth holds up the entrails in the direction of a severe figure in a toga.

This image epitomizes the disintegration of the formalized anatomical

---

39. Oribasius (325–403) and Paulus Aegineta (625–690) are two of the most important channels for the diffusion of Galen's medicine, and they contributed greatly to his establishment in the Middle Ages as the undisputed medical authority. For a detailed discussion of the role of the late antique and Byzantine compilers in the history of medicine, see M. Neuberger, *Geschichte der Medizin* (Stuttgart, 1906), 2:104–28.

quodlibetum: students touching the cadaver and rummaging around its innards; the mixing, if not actually the superimposing, at a spatial level, of theory with the practice of anatomy; the absence of the book and of a person reading or of one who possesses an obviously commanding role in the action; and the overturning of hierarchical relations exemplified by the disposition of the figures around the body being demonstrated. To this should be added the animation of the discussions and the vehemence of the gesturing, which mark a decided break with the composed formalism of the scenes depicted on title pages previously considered.

This edition of the *De anatomicis administrationibus* was prepared entirely in Paris from translation to publication, leading one to suppose that the anatomical demonstration it portrays reflects the actual manner in which it was performed at Winther's university. In the preface of the *De humani corporis fabrica,* Andreas Vesalius, who had been Winther's pupil, relates a personal episode from the years he had studied in Paris that helps to contextualize and evaluate this unusual image. The medical students at the university were entitled to attend no more than two dissections during the entire period of their enrollment. These were performed inadequately and superficially by unskilled barbers (*imperitissimi tonsores*) at public anatomy lessons. Vesalius had carried out some animal dissections under the direction of Jacques Du Bois (Iacobus Sylvius), his teacher along with Winther. Once, at the urging of his companions and preceptors, he succeeded not only in participating in a third anatomy exercise but in actually doing the dissecting himself. In this case he was given greater freedom than was usual. On another occasion, after the barbers had been dispensed with, Vesalius was able to demonstrate publicly how the muscles of the hand in a dissected cadaver were constructed.[40] In Paris, according to Vesalius, it appears that the anatomy lesson was structured strictly along the quodlibetarian model, namely as it was depicted on the title pages of Mondino's and Berengario's writings. The events described by Vesalius constitute exceptions to the rule, but the frequency with which they occurred is difficult to estimate. The title page of

40. "Verum id studium [of anatomy] neutiquam successisset, si quum Parisiis medicinae operam darem, huic negocio manus non admovissem ipse, ac obiter mihi et consodalibus ab imperitissimis tonsoribus in una atque altera publica sectione visceribus aliquot superficietenus ostensis acquievissem. Adeo enim perfunctorie illic, ubi primum medicinam prospere renasci vidimus, Anatome tractabatur, ut ipse in brutorum aliquot sectionibus sub celebri ac nunquam satis laudato viro Iacobo Sylvio versatus, tertiam cui unquam mihi adesse obtigit sectionem, solito absolutius, et sodalium et praeceptorum hortatu adductus publice administrarem. Quum autem secundo (tonsoribus ab opere iam relegatis) illam aggrederer, musculos manus cum accuratiori viscerum dissectione conatus sum ostendere (A. Vesalius, *De humani corporis fabrica* [Basel: J. Oporino, 1543], fol. 3r).

the *De anatomicis administrationibus* reproduces one of those sporadic cases in which certain students were able to dissect the cadaver with their own hands.

This image provides evidence of a change in process. Although the instructors keep their (much reduced) distance from the dissecting table and are not yet touching the cadaver with their own hands, the fact that medical students are represented in the act of groping inside the abdomen is symptomatic of a new attitude. The dichotomy between theory and practice in both the teaching and learning of anatomy is evidenced in the title pages examined thus far, as in all representations of dissection. Since antiquity the texts had constantly insisted on the necessity of instructing students not only in anatomical theory but also in the practice of dissection. Students acquired such technical experience, as Vesalius himself recalls, by practicing on animal corpses, and certainly not on human cadavers, that were used only according to university regulations (which will be the subject of the following chapter) on the occasion of public anatomy lectures. These lectures were celebrated according to a strictly formalized ritual once or twice yearly. They clearly separated out the theory and practice of anatomy as it regarded both the roles the celebrants played in the ritual and the space they occupied. To substitute a student for the ignorant barber in the role of *sector,* as the title page of the *De anatomicis administrationibus* does, implies that there is a shift in the practical aspect of anatomical knowledge taught by certain texts that the ritual organization had previously relegated to the hands of barbers or surgeons. In this way manual dexterity, which was previously only implied and was frequently disparaged, was reunited in the person of the student to theoretical knowledge of which it was, at the same time, both fruit and indispensable complement.

Research on the iconography of the anatomical lecture has maintained that this title page constitutes an isolated example, but it is precisely this exceptional feature that marks the moment of transition from a quodlibetarian model to the one inaugurated in the title page of the *De humani corporis fabrica* by Vesalius.[41] However, this type of dissecting tableau would not be adopted in the subsequent iconography. It is nevertheless far from being a unique example of it. A series of images produced between the final decades of the fifteenth century and the early ones of the sixteenth provides a model for the author of the woodcut appearing in Simon de Colines's edition of the *De anatomicis administrationibus.* I shall try to explain this unusual iconographic organization.

41. See, for example, W. Heckscher, *Rembrandt's Anatomy,* especially pp. 52–57.

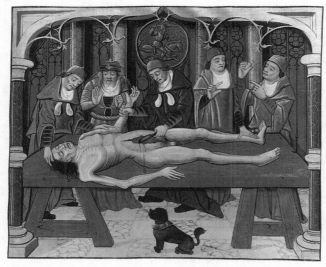

Figure 12   Bartholomaeus Anglicus, *Le propriétaire des choses*. (Bibliothèque Nationale, Paris). A dissection scene in an illuminated French manuscript from the end of the fifteenth century, Bartholomaeus Anglicus's encyclopedic *Le propriétaire des choses*. Although the work was not specifically written for physicians or medical students, sections of it are dedicated to the human body. As Galen himself wrote in the *De anatomicis administrationibus*, knowledge of the body, and even of its parts, is not just the prerogative of physicians, but is also, even if not especially, that of philosophers. The concept of anatomy as a body of knowledge intended as much for philosophers as for physicians is taken up again in the Renaissance by such authors as Berengario da Carpi and Andreas Vesalius. On the other hand, a series of broadsheets from the sixteenth and seventeenth centuries and certain texts intended for the lay reader that were circulated by the popular press of the ancien régime prove that knowledge of the human body constituted a discipline that, appropriately re-adapted, crossed the borders of medical use, educated or elitist.

The first is an illumination from the final quarter of the fifteenth century in a French manuscript of the encyclopedia by Bartholomaeus Anglicus, *Le propriétaire des choses* (fig. 12).[42] It contains an illustration of the dissection of a body on which the customary incision from the pubis to the sternum has been made. Five people are placed along one side of the table. The figure at the center, holding the wrist of the cadaver with his right hand, physically carries out the dissection with a knife. On his left two other physicians, dressed identically, are discussing the dimensions of some part of the body intently; one of them is touching the ankle. The head and an arm are supported by a person whose garments pair him

42. Paris, Bibliothèque Nationale, MS. fr. 218, fol. 56r. It was mentioned first in A. M. Cetto, "Die Sektion in der mittelalterlichen Miniatur," *Ciba Symposium* 5 (1957): 168–72. Cetto's article also appears simultaneously in English: "Dissection in Miniatures of the Middle Ages." Cf. G. Wolf-Heidegger and A. M. Cetto, *Die anatomische Sektion*, n. 51.

with the figure acting as the *sector,* while an older person, dressed in more solemn and elegant clothing, carefully observes the event on the anatomical table, perhaps advising on correct procedure. In this miniature it is not clear whether the representation is of an anatomy lesson or only of a practical exercise on the cadaver (an autopsy, for example). It is also difficult to establish with any certainty who is actually coordinating the operation and who the students and the teachers or teacher might be. This equivocal image of anatomical practice is somewhat perplexing, but we can only look at the evidence on the page. The toga worn by the person dissecting the cadaver makes it clear that he is not a barber. It seems plausible, instead, that he, along with the three other figures, is represented as beardless to indicate that he is a student, while the older person could be identified as a teacher.

This image is contained in a manuscript that was undoubtedly not intended for a specialized audience, and the author of the miniature certainly did not wish to produce an image that corresponded realistically to dissection as it was practiced. Despite this, it is similar to a series of woodcuts contained in a number of printed editions of Bartholomaeus Anglicus's *Le propriétaire des choses,* all published toward the end of the fifteenth century, which is presumably when the manuscript was written. The fifth book of the 1482 Lyons edition, entitled *De lomme et de ses parties. Le premier chapitre du corps de lomme,* opens with a woodcut containing a representation of the anatomy lesson that resembles that of the miniature (fig. 13).[43]

Since we do not have a precise date for the latter, it is impossible to establish the sequence of iconographic borrowings. Here, too, there are five figures, one of whom is delving with his hands into a cadaver's abdominal cavity. All are dressed identically and possess rather youthful features, with the exception of the figure at the feet of the body, who seems more mature and is perhaps intended to represent the teacher, even if in a rather ambiguous manner. In the images contained in successive editions of this encyclopedia, the figures depicted are portrayed in ways that increasingly reveal the role and status of each of them more clearly. Though the figure of the teacher directing the anatomy session is hard to

43. Bartholomaeus Anglicus, *Le propriétaire des choses* (Lyons: Matthias Husz, 1482) (HC, n. 2514). On this edition, see A. M. Hind, *An Introduction to a History of Woodcut,* 2 vols., (London, 1935), 2:601. A reproduction of this woodcut is included in the (Lyons, 1485) edition of this work (HC, n. 2518). For the anatomical woodcuts in the encyclopedia by Bartholomaeus Anglicus (figs. 13–17), see G. Wolf-Heidegger and A. M. Cetto, *Die anatomische Sektion,* nn. 50–57.

distinguish in these first woodcuts, it becomes better defined later by his clothing,[44] while his way of gesturing becomes increasingly explicit.[45] Only in two Parisian editions of this work, one published between 1493 and 1500 (fig. 16),[46] and the other published in 1510 (fig. 17),[47] does the figure of the instructor delivering the anatomy lecture assume all the characteristics of clothing and gesture appropriate to his position: the garments are of a more distinctly academic style and differ from those of the other bystanders; he is portrayed in the unambiguous act of discussing and explaining; he does not touch the cadaver, which is instead handled by a student; and the knife used to make the incision now lies abandoned on the table.

In 1498 this very type of illustration was used for the dissection scene in a French edition of the surgical treatise by Guy de Chauliac.[48] The image, in which the student assumes the functions of a *sector,* opening up the cadaver in the presence of his colleagues and under the supervision of an instructor who offers an explanation on the parts of the human body without the benefit of an anatomical text, is now to be found in strictly medical publications.

This series of images probably provided the model that guided the author of the anatomy disputation reproduced on the title page of the *De anatomicis administrationibus.* It is not by chance that this representation was initially produced and conceived for a nonacademic text such as the *De proprietatibus rerum,* and it demonstrates the author's summary knowledge of the canonical ways in which the anatomy lesson was conducted and represented. Strict adherence to the rules was certainly irrelevant in an image that merely had to suggest to a public of nonspecial-

44. See Bartholomaeus Anglicus, *El libro de proprietatibus rerum* (Toulouse: Henrique Meyer d'Alemaña, 1494 (IGI, n. 1261), where one can make out a person with a large-brimmed hat (fig. 14).

45. In the edition published at Lyons by Jean Jenin le Dyamantier in 1500 (HC, n. 2519), two persons are wearing ermine, one is conversing, and the other has his hands in the abdomen of the cadaver (fig. 15). The same model is used in the anatomical engraving published in the Latin version of Bartholomaeus's text: *De proprietatibus rerum* (London: Wynkyn de Worde, c. 1495?) (HC, n. 2520).

46. Bartholomaeus Anglicus, *Le propriétaire des choses* (Paris: Antoine Verard, [n.d.]) (HC, n. 2512).

47. Bartholomaeus Anglicus, *Le propriétaire des choses* (Paris: J. Petit et M. Le Noir, 1510).

48. Guy de Chauliac, *Le livre appelé Guidon. De la pratique en chirurgie* (Lyons: Jehan de Vingle, 1498) (HC, n. 4815b). A copy of this extremely rare incunable, perhaps the only one, is preserved at the Bibliothèque Municipale of Besançon (cf. G. Wolf-Heidegger and A. M. Cetto, *Die anatomische Sektion*, n. 117).

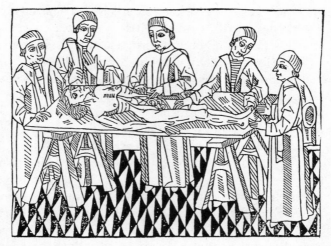

Figures 13–15   Figure 13: Bartholomaeus Anglicus, *Le propriétaire des choses* (Lyons: Matthias Husz, 1482). Figure 14: Bartholomaeus Anglicus, *El Libro de proprietatibus rerum* (Toulose: Henrique Meyer d'Alemaña, 1494; Biblioteca Marucelliana, Florence). Figure 15: Bartholomaeus Anglicus, *Le propriétaire des choses* (Lyons: Jean Jenin le Dyamantier, 1500). Anglicus's *Le propriétaire des choses* has frequently appeared in print, in Latin and in the vernaculars, especially in France and Britain. An engraved image of the dissection scene is contained in many of these editions. In these representations it is difficult to discern the specific roles performed by the individual figures, to establish a hierarchy or to attribute a specific function to them. The book is not addressed to a specialized public, so the intention of the executors of these images—all of which were clearly inspired by the same model and which are related to one another—is simply to guarantee that the theme of the chapter to which they refer be easily recognizable: a cadaver with its abdomen opened from the sternum to the pubis stretched out on a wooden table and a group of persons gathered around it is evidence enough to suggest to the reader that the subject of the chapter they are reading concerns "Man and his parts."

ists, without entering into too many details, the subject of one of the books contained in the encyclopedia. Nevertheless these images, while remaining close to the original model, reveal a progressive transformation and a more precise determination of the roles played by the celebrants, which suggests a parallel attempt by the woodcutters to adapt it for the purpose of achieving a greater adherence to reality. While the subject depicted in the lower portion of the scene in the Galen anatomy of 1531 refers to these exceptional cases in which students open the cadaver, these unusual images of dissection provide an influential iconographic model at the same time.

The illustrated title page enhances the work while providing information to the reader. It makes a text instantly recognizable. A table with a dissected cadaver on it and figures standing about it implies that the book

is an anatomical treatise or at least a medical text. This is the essential aim of the publisher, the editor, and particularly the printmaker or designer of the illustration. From this perspective, it is hardly surprising if the iconographic models just considered assumed a greater importance in the production of images than did the actual subject itself. This seems to have been the case for the artist of the woodcut that appears at the beginning of the Latin translation of Galen's anatomy published by Simon de Colines. A different purpose animates the title page of Vesalius's text, however, in which the depicted scene seems to have a function that goes well beyond the simple revelation of the book's content.

Figures 16–17   Figure 16: Bartholomaeus Anglicus, *Le propriétaire des choses* (Paris: Antoine Verard, [1493–1500]; Bibliothèque Nationale, Paris). Figure 17: Bartholomaeus Anglicus, *Le propriétaire des choses* (Paris: J. Petit and M. Le Noir, 1510). Certain versions of the woodcut representing the dissection scene used in two of the Parisian editions of *Le propriétaire des choses* by Bartholomaeus Anglicus suggest the introduction of a hierarchical arrangement in the roles of the persons present. In these two images one can tell by his dress and gestures who is orchestrating the anatomical demonstration. Here it might be possible to discern an iconographic model of the anatomy lesson—different from the formalized quodlibetarian one—produced by the Parisian publishers to illustrate Bartholomaeus's encyclopedia, and then adapted for the title pages of the Latin translation of Johann Winther's *De anatomicis administrationibus* (fig. 11). This model is distinguished by the absence of a separation or a contraposition between the space for reading and the space for dissection. But an element that remains common to the quodlibetarian and Parisian models is that the instructor—or at least the person who appears to be in the act of orally discussing and describing the dissection—is keeping his distance from the cadaver and is not touching it.

THE SHIFT: THE TITLE PAGE OF ANDREAS VESALIUS'S "DE HUMANI CORPORIS FABRICA"

In 1543 Andreas Vesalius, a professor at the University of Padua medical school, published his *De humani corporis fabrica libri septem* and, contemporaneously, two editions, one in Latin and one in German, of the *Epitome,* a sort of brief compendium of the larger work.[49] These volumes, published by Johannes Oporinus in Basel, constitute an event of extraordinary importance both for the history of anatomy and for the history of the book. By directly observing the cadaver and through his possession of a profound knowledge of earlier anatomical literature, Vesalius was able to confirm, discuss, and correct everything that had been said previously about the different parts of the body. He often entered into open controversy with those who fiercely defended a still vigorous Galenic tradition. His repeated references to dissections he had performed provide what the author himself suggested was the major feature of the Vesalian revolution as later defined by historians.

These developments, crucial to both the textual and technical nature of the anatomical discipline, were tied to a series of insights related to the book trade, which make this publishing venture by Vesalius and Oporinus one of the most important and astute successes of the first century of printing. One of Vesalius's greatest merits lay in the fact that he was the first to have understood and exploited this still novel method of expressing and communicating knowledge to its full potential. The publication of the *Epitomi,* the use of splendid and detailed illustrations as visual aids, the technique of relating text to images, the subdivisions into books, chapters and paragraphs, the illuminated initials that open each of these sections, the title page and portrait of the author—each of these features was thought out and realized by Vesalius, and his collaborators, as an aspect of the whole enterprise; this kind of integral arrangement had no real precedent in previous publications and conventions and often openly clashed with them. The *Fabrica* gives a newfound coherence to preex-

49. A. Vesalius, *De humani corporis fabrica libri septem* (Basel: J. Oporinus, 1543); A. Vesalius, *De humani corporis fabrica Epitome* (Basel: J. Oporinus, 1543); A. Vesalius, *Von des menschen Cörpers Anatomey, ein kurtzer aber vast nützer Ausszug auss D. Andr. Vesalii . . . Bücheren von ihm selbs in Latein beschriben unnd durch D. Albanum Torinum verdolmetscht. . . .* (Basel: gedruckt bey Johann Herpst, genant Oporino, 1543). Vesalius was born in Brussels in 1514 and descended from a long line of physicians. After attending the universities in Louvain and then in Paris, he completed his medical studies in Padua in 1537. That same year he was entrusted with the chair of surgery (comprising also the teaching of anatomy) which he kept until 1542. It was in Padua that he conceived and wrote the *Fabrica* and the *Epitome.*

isting material. The so-called Vesalian revolution was probably nothing but a transformation of the forms and ways of using certain tools (dissection and printing, to mention only the most obvious) and of already acquired knowledge.

The practice of dissection and the necessity of direct observation in the study and teaching of human anatomy were recommended by Vesalius as essential tools for the anatomist. This was not new. Berengario da Carpi, Alessandro Benedetti, Niccolò Massa, Johann Winther, and even Galen (if we go back to antiquity), had insisted on this.[50] Of course, some pre-Vesalian anatomists had the opportunity to open cadavers with their own hands and to observe them with their own eyes, but this was never part of the public anatomy lesson, which remained tied to the formalized ritual of the academic tradition sanctioned by the university statutes. If we consider that the lessons, at least on paper, on the basis of these rules, were supposed to be the only occasions during which the dissection of the human body was permitted, Vesalius's harsh criticism of this practice, which was useless and which detracted from its didactic and research purpose, is entirely justified.

Vesalius traces the custom of keeping theory and practice separate in medicine, and especially in anatomy, back to the time of the Gothic invasions, "when all sciences came to ruin."[51] This, he wrote in the preface, was caused by the discredit into which manual labor, (*manus opera*), had fallen. In response to this development, physicians from that time onward began to occupy themselves solely with the prescription of medicines and diets, rejecting other aspects of medical practice; these were left to surgeons, considered simple attendants (*famuli*). As a result, that primary and most ancient branch of medicine that consisted primarily of an investigation into nature (*naturae speculatio*), was pushed to one side.[52] This custom survived, wrote Vesalius, until the sixteenth century: "in medicine today teachers of healing, remove themselves from physical contact, as from the plague." This attitude had disastrous consequences: "not only has true knowledge of the viscera by physicians suffered, but the very art of dissection has been totally ruined."[53]

50. See chapter 4.
51. A. Vesalius, *De humani corporis fabrica*, fol. 2r.
52. Ibid, fol. 3r: "Atque ita temporibus successu, curandi ratio tam misere divulsa est, ut medici quidem, se physicorum nomine venditantes, medicamentorum et victus ad reconditos affectus praescriptionem sibi duntaxat arrogaverint: reliquam autem medicinam, iis quos Chirurgos nominant, vixque famulorum loco habent, relegarint, turpiter a se quod praecipuum et antiquissimum est medicinae membrum, quodque naturae speculationi (si modo quod aliud) in primis innititur, depellentes."
53. Ibid, fol. 2v.

These considerations of the history and the state of medical practice served as a preamble to Vesalius's vehement polemic against the ritual of the anatomy lesson performed (where it was performed)[54] according to the quodlibetarian model represented in the title pages examined:

> It has gone so far that these people [the barbers] even kept for themselves the very difficult and secret art [anatomy] entrusted to them and this disastrous loss of the therapeutic branch has also introduced a detestable habit in the *gymnasia,* by which [the barbers] practice dissections on human bodies while [the physicians] describe the parts of the body. Indeed these latter have no experience at all of the business, but having learned by heart matters from others' books they display their learning by croaking with much ado from a high pulpit. The barbers, so unaccustomed to speaking, cannot explain the dissections to the public and so ruin what is necessary to demonstrate following the orders of the physician. The latter, unaccustomed to practicing dissections on the human body, with his book arrogantly plays at steering the affair. As in this way everything is badly taught at the *scolae* and time is wasted in ridiculous discussions, in this confusion, the public is shown less than that which a butcher could teach a physician at the market.[55]

In opposition to this manner of conducting the anatomy lesson, Vesalius would himself dissect cadavers during the public demonstrations, and he frequently invited his students to perform dissections, since he believed that doing so was didactically important to the education of a good physician. This also was not new: other anatomists before Vesalius had personally performed dissections, and Vesalius himself, while still a student, had done so on at least two occasions in Paris. It would appear, however, that the performances by Vesalius's students did not take place during the annual public demonstrations held in Padua, but in private sessions. These were a sort of secret lesson, not included in the *curricula,* in which Vesalius (and other anatomists after him) privately trained some of the students in the study of the human body. Similarly, pre-Vesalian anatomists never personally performed dissections in a public session. Given the scarcity of bodies on which this type of research could be exercised, the dissections were either carried out privately on cadavers obtained out-

54. In many universities the medical *curricula* still did not provide for dissection in the teaching of anatomy: "ut aliquot Gymnasia praeteream, ubi de humani corporis compage resecanda vix unquam cogitatur" (ibid, fol. 3r).

55. Ibid. [The translation is taken from A. Carlino, "The Book, the Body, the Scalpel," RES 16 (1988): 32–50, at p. 39, trs. note].

side official channels, or they were true and proper autopsies conducted in hospitals where anatomical investigation was associated with the study of disease and the causes of death.[56]

In the title page of the *Fabrica*, (fig. 18),[57] instead, Vesalius is portrayed in the act of dissecting a cadaver during a public anatomy lesson. This image requires careful examination since it became the prototype of the prevailing iconographic model of a public anatomy lesson for over a century. The artist who conceived the design and the engraver of the image—one of the masterpieces of sixteenth-century xylography—are unknown, though many hypothetical attributions have been suggested. The names that have been proffered most often are those of Domenico Campagnola and Jan Stephan von Calcar, a Flemish painter working in Venice who had already collaborated with Vesalius in the realization and the financing of the *Tabulae Anatomicae Sex*[58] and who, moreover, is the author of the latter's portrait contained in the *Fabrica* as well as at the end of the *Epitome* (fig. 19).[59] Aside from the problems connected with the attribu-

56. To cite just two examples, Alessandro Benedetti and Niccolò Massa frequently recall dissections taking place in hospitals for autoptic purposes. See by the two writers, respectively, *Anatomice sive historia corporis humani. Ejusdem collectiones medicinales seu aphorismi* (Venice: Bernardino Guerraldo Vercellensis, 1502), and *Liber introductorius anatomicae sive dissectionis corporis humani, nunc primum ab ipso auctore in luce aeditus in quo quamplurima membra operationes et utilitates tam ab antiquis quam a modernis praetermissa manifestantur* (Venice: Francisci Bindoni and Maphei Pasini, 1536).

57. The same illustration is printed in the two editions of the 1543 *Epitome*.

58. These are six illustrated broadsides, three designed by Vesalius and three by Calcar, published in Venice in 1538. See C. Singer and R. Rabin, *A Prelude to Modern Science: Being a Discussion of the History, Sources and Circumstances of the "Tabulae anatomicae sex" of Vesalius* (Cambridge, 1946). The *Tabulae* are not the first examples of anatomical sheets. This was actually a rather popular typographical genre between the end of the fifteenth century and the beginning of the seventeenth. There are two principal collections of these, the first at the Medical Center Library of the University of Michigan, the other at the Wellcome Library, London. Many other copies are dispersed in European libraries, and very few have been inventoried. A systematic cataloguing of these holdings would be highly desirable. There are a number of studies—now somewhat dated, occasionally confused, and never complete—devoted to this material. The most important articles on these anatomical sheets are: L. Crummer, "Early Anatomical Fugitive Sheets," *Annals of Medical History* 5 (1923): 189–209; L. Crummer, "Further Information on Early Anatomical Fugitive Sheets," *Annals of Medical History* 7 (1925): 1–5; J. G. De Lint, "Fugitive Anatomical Sheets," *Janus* 28 (1924): 78–91; L. H. Wells, "A Remarkable Pair of Anatomical Fugitive Sheets," *Bulletin of the History of Medicine* 37 (1964): 470–76. On anatomical fugitive sheets see A. Carlino, *Paper Bodies: A Catalogue of Anatomical Fugitive Sheets in the Age of Printing and Dissecting* (London, 1999).

59. On the attribution of the title page and illustrations of the *Fabrica*, see: M. H. Spielmann, *The Iconography of Andreas Vesalius* (London, 1925), pp. 124–39; L. Crummer, "An Original Drawing of the Title Page of Vesalius' Fabrica," *Annals of Medical History,* n.s. 2 (1930): 20–30; H. W. Cushing, *A Bio-Bibliography of Andreas Vesalius* (New York, 1943); L. Choulant, *History and Bibliography;* J. B. de C. M. Saunders and C. D. O'Malley,

tion of the woodcut and of all the illustrations contained in this treatise, what is of most interest is the fact that Vesalius personally supervised the work and gave substantial advice to the artist, whoever he might have been.

The title page thus assumes the character of a true and proper manifesto of the concept of anatomy and the practice of dissection championed by Vesalius. The reader of the *Fabrica,* from the very beginning, is conscious that this is a work radically different from any previously produced on these subjects.

The title page of the *Fabrica* is a rather elaborate and complex woodcut. Though some of its elements have a solid basis in reality, others seem to have been modified by the author's aesthetic or rhetorical values;[60] others still are of an eminently symbolic character and serve to make the scene represented recognizable and therefore more successful in its purpose than would have been the case in a realistic representation. The meaning of some of the symbolic elements, such as, for example, the significance of the nude man clinging to the columns at the right, is still under debate.[61] Without venturing to offer either a hypothetical or a conjectural reading of the hermetic and therefore elusive implications of this

---

The Anatomical Drawings of Andreas Vesalius (New York, 1950); W. Artelt, "Das Titelbild zur *Fabrica* Vesals und seine kunstgeschichtlichen Voraussetzungen," *Centaurus* 1 (1950–51): 66–77; W. M. Ivins Jr., "What about the Fabrica of Vesalius?," in S. W. Lambert, W. Wiegland, and W. M. Ivins, *Three Vesalian Essays to Accompany the Icones Anatomicae of 1934* (New York, 1952), pp. 43 ff.; L. Premuda, *Storia dell'iconografia anatomica* (Milan, 1957), pp. 111 ff.; A. M. Cetto, "Zwei unbekannte Darstellungen des Andreas Vesalius (Spicilegium anatomicum et Vesalium)," in *Verhandlungen. XIX internationaler Kongress für Geschichte der Medizin 1964* (New York and Basel, 1966); F. Guerra, "The Identity of the Artists Involved in Vesalius' Fabrica 1543," *Medical History* 13 (1969): 37–50; M. Kemp, "A Drawing for the Fabrica and Some Thoughts upon the Vesalius Muscle-Men," *Medical History* 14 (1970): 277–88; M. Muraro and D. Rosand, *Tiziano e la silografia veneziana del Cinquecento* (Vicenza, 1976), pp. 123 ff.; C. M. Bernstein, "Titian and the Anatomy of Vesalius," *Bollettino dei Musei Civici Veneziani* 22 (1977): 39–49.

60. Here I use "author," to mean the collective author–that is, designer, engraver, Vesalius, and Oporinus.

61. On the opposite side of the title page, a figure of the clothed youth is juxtaposed to this figure. Both constitute an enigma. None of the interpretations suggested thus far seems convincing and proven: M. H. Spielmann (*The Iconography,* p. 135) hypothesizes that the naked man symbolizes "study from nature and study from conventional teaching—by book alone," while the clothed figure could represent "a soldier whose body may one day find its way to the dissecting table." C. D. O'Malley (*Andreas Vesalius of Brussels, 1514–1564* [Berkeley and Los Angeles, 1965], p. 141) maintains that the nude stands for superficial anatomy and "may be expressing horror and loathing of the fate before it as it observes the dissection in progress." W. Artelt ("Das Titelbild," p. 66) and R. Herrlinger, *History of Medical Illustration,* pp. 103–4) consider them, more prudently, as recurring elements in the contemporary woodcut representations of theatrical scenes.

Figure 18   Andreas Vesalius, *De humani corporis fabrica libri septem* (Basel: Johannes Oporinus, 1543). Andreas Vesalius's *Fabrica* constitutes a watershed in the history of modern anatomy. Historiography has underlined, perhaps too emphatically, the "revolutionary" character of this work: the periodization in anatomical publishing is denoted either as "pre-Vesalian" or "post-Vesalian." Vesalius himself, well aware of the innovative character of his treatise and the possibilities for self-promotion provided by the printing press, set himself apart from the earlier anatomical tradition beginning with the book's title page itself, which is a sort of iconic manifesto of the new anatomy. In significant contrast to the previous iconographic tradition, Vesalius—the only figure to direct the anatomical demonstration both in its theoretical and in its practical aspects—has represented himself in the act of dissecting the cadaver of a woman while surrounded by an agitated but attentive composite group. Clerics, laymen, students, and the simply curious lean out from the platforms and archways to gain a better view, talk to one another, leaf through books to compare what they are seeing with what they know, feverishly take notes, and more or less discretely gawk at the proceedings from behind columns and from the windows above. Vesalius sardonically represents two barbers—the *sectores* of the quodlibetarian model—here by placing them at the feet of the dissecting table where they argue over who will sharpen the knives and pass them to the anatomist. The scene is dominated by a skeleton clutching a magisterial baton: this is undoubtedly a reference to the theme of *memento mori* and to the triumph of death, which recurs as a metaphor in anatomical iconography both before and after Vesalius, but it is also an allusion to the importance assigned by Vesalius to the study of osteology, to the weight-bearing structure of the fabric of the human body.

Figure 19   Andreas Vesalius, *De humani corporis fabrica libri septem* (Basel: Johannes Oporinus, 1543). In the portrait contained in the *Fabrica* as well as in the two editions of the *Epitome*—a brief synthesis of the larger work that Vesalius published in Latin and in German translation, also in 1543—the author restores a key concept of his approach to anatomy and his notion of the professional profile of the anatomist-physician. Vesalius shows himself in the act of dissecting the forearm of a disproportionately large woman. In particular he highlights the flexing muscles of the fingers. The instruments unique to the anatomist, representing practical skill and intellectual competence, are arranged on the table: on one side the knives and the surgical instruments, on the other a pen, an ink well, and a sheet of paper. Polemically, Vesalius is also trying to emphasize that, contrary to what occurred in the anatomy demonstrations organized according to the formalized rite of the *quodlibetum,* it is the direct observation of the body and of its parts that produces the anatomical text.

title page, I will consider an analysis of the complete structure of the scene and will limit myself to those characteristics that the preface and the text of the *Fabrica* permit us to decipher.

The public dissection is taking place in the open, inside a wooden theater that stands beneath a semicircular frame. It has been suggested that this is intended to represent the courtyard of the Bo', the University of Padua, in which the same Doric decorations of the trabeation in the form of bucrania, triglyphs, and vases could be found.[62] This type of decoration is actually encountered with great frequency in sixteenth-century Italian, especially Venetian, architecture. It is described in minute detail and abundantly illustrated in Sebastiano Serlio's *Regole generali di architettura*, one of the most successful and widely circulated architectural texts of the first half of the century.[63] Serlio is connected to the title page of the *Fabrica* in another respect as well: he is the planner and builder of a temporary wooden theater, of the type represented on the title page, which he constructed for a special occasion in the courtyard of a Venetian palace.[64] The theater depicted in the woodcut is in fact a temporary structure that was erected every time an anatomy lesson was held. Vesalius himself notes that theaters similar to the one on the title page were prepared for such occasions, both in Bologna and in Padua.[65]

A packed audience of seventy or eighty people crowds the space and peers from behind the columns and from the windows above the theater.[66] The dissection taking place on the table at the center has captured the attention of the majority of the bystanders, while a few others are talking busily. From their dress and attitudes, we can distinguish students, professors, clerics, physicians, and even common people, simple curiosity seekers who have come to observe the anatomy lesson. Anatomical knowledge is becoming accessible to a broader public; an interest in discovering the organization of the human body has developed beyond the borders of the university, and the academic rigidity seen in older title pages has been transformed into this lively scene.

62. See M. H. Spielmann, *Iconography,* p. 129; O'Malley, *Andreas Vesalius,* pp. 140–41; M. Muraro and D. Rosand, *Tiziano,* p. 127.

63. S. Serlio, *Regole generali di architettura . . . sopra le cinque maniere degli edifici, cioè Toscano, Dorico, Jonico, Corintio e Composito con gli esempi delle antichità, che per la maggior parte concordano con la dottrina di Vitruvio* (Venice: F. Marcolini, 1537). This is the fourth book of Sebastiano Serlio's architectural treatise in seven books. For the troubled publishing history and a careful discussion of this work, see W. B. Dinsmoor, "The Literary Remains of Sebastiano Serlio," *The Art Bulletin* 24 (1942): 55–91. For Serlio's influence on contemporary art, see C. Gould, "Sebastiano Serlio and Venetian Painting," *Journal of the Warburg and Courtauld Institutes* 25 (1962): 58–64.

64. S. Serlio, *Il secondo libro d'architettura* (Paris: Iehan Barbé, 1545). This book, focusing on perspective and published together with *Il primo libro d'architettura,* which deals with geometry, contains a brief *Trattato sopra le scene* in which the description of the pit closely resembles the theater of Vesalius. This is where the author refers to the wooden theater built in Vicenza (fols. 43v & 44r).

65. A. Vesalius, *De humani corporis fabrica,* p. 548.

66. At least five hundred people were present at the anatomical demonstration performed by Vesalius in Bologna in January, 1540, in collaboration with Matteo Corti, who read

At the center of the page a skeleton holds a magisterial cane. This is certainly another of those symbolic elements mentioned earlier for which it is difficult to offer a concrete and convincing explanation. On the one hand, it may be there to stress the importance of osteology for the anatomist. Vesalius frequently insisted that a skeleton be handy during dissections, and he mentions the reconstruction of the human frame as one of the most useful exercises in the education of the young physician.[67] He relates that he had himself assembled at least three human frames for the teaching of medicine in general and of anatomy in particular: in 1536 he put one together in Louvain,[68] in 1538 he did the same with the bones of a youth in Padua,[69] and in 1543 he donated one to the University of Basel on the occasion of his visit to that city for the printing of his anatomical treatise. On the other hand, the skeleton is also the symbol of death. The 1555 edition of the *Fabrica* (fig. 20) has a title page that was altered in a few details in respect to the 1543 edition.[70] One of these is the scepter-like rod that becomes, in the reworking, a scythe, thereby emphasizing the portentous symbolism of the skeleton. Numerous illustrated title pages of the sixteenth and seventeenth century, following the Vesalian example, include a skeleton among the prominent iconographic elements and increasingly allude to death.[71]

---

from Mondino's *Anatomia*. This was related by Baldasar Heseler, a student who attended that cycle of lessons and whose detailed manuscript notes, preserved at the Royal Library in Stockholm, have now been published: B. Heseler, *Andreas Vesalius' First Public Anatomy at Bologna, 1540*, ed. R. Eriksson (Uppsala, 1959). These notes differ considerably from those taken by Vitus Tritonius Athesinus (preserved in the National Library, Vienna, Cod. 11.195; med. 119) during Vesalius's Paduan lectures in 1537 and expanded subsequently with the assistance of anatomical writings by Galen and Winther.

67. See, for example, Bk. I, chapter 39 of the *De humani corporis fabrica*, pp. 155 ff., in which Vesalius explains how a skeleton is prepared. And at p. 548, alluding to the theater represented in the title page, he writes, " . . . leviter praesenti sceleto praefaberis vix enim hodie ulla est Academia, quae non unum atque alterum habeat, quod sectioni semper utilissime adhibeatur." Note also the bones resting on the shelf below the dissecting table, which may themselves be an allusion to the importance of osteology.

68. Ibid, p. 161.

69. In the introduction to the *anatomicae sex*, Vesalius writes that the three tables representing the skeletal structure were drawn from life by Calcar, who copied "meum *skeletoy* nuper in studiosorum gratiam constructum" (*Tabulae*, table 1. See the republication by Singer and Rabin, *A Prelude to Modern Science*, p. 1 and note 8.)

70. A. Vesalius, *De humani corporis fabrica libri septem* (Basel: J. Oporinus, 1555). At the same time as the *Fabrica*, even the *Epitome* was reissued in Latin by the same publisher.

71. In an engraving dating from 1610 representing the anatomical theater in Leiden, skeletons of men and animals take the place of spectators, and many of them are holding a sort of pennant with slogans concerned with death and the frailty of life. See E. Höllander, *Die Medizin in der klassischen Malerei* (Stuttgart, 1923), p. 53; W. Brockbank, "Old Anatomical Theaters and What Took Place Therein," *Medical History* 12 (1968): 371–84; L. Wilson, "William Harvey's 'Praelectiones:' The Performance of the Body in the Renaissance

Before considering what is taking place on the dissecting table at the center of the scene, I should like to examine three details of the title page that appear to be rather significant and that are decipherable through the preface of the *Fabrica:*

(1) Two figures in rather humble dress at the foot of the dissecting table are arguing over a razor. They are barbers now polemically relegated to a secondary unimportant role: that of sharpening the blades used by the anatomist in his dissection. Barbers are one of Vesalius's principal targets in the controversy that he vehemently wages in the preface against the anatomical lesson and the associated dissecting practice conducted according to the quodlibetarian model: the *tonsores* (barbers) are defined as "highly unskilled," "scarcely knowledgeable plebeians and servants in the medicinal art," and, again, as "unlearned."[72]

(2) In the right foreground a man, perhaps an attendant, is shown with a barking dog.[73] An austere, classically garbed person with a white beard remonstrates with a few agitated spectators, inviting them, while gesturing with his left hand, to concentrate on the event at the center of the scene, where the dissection is taking place, where knowledge is being created. It could be that this gesture is intended to represent the ancient medical tradition that sees Vesalius as an innovator and as the heir of classical science, and the empirical examination of the cadaver as the essential instrument of a newly reestablished anatomy. The figure has frequently been identified with Galen, and this interpretation is convincingly corroborated by numerous references to him in the text. Galen, "who never cut up a human body," writes Vesalius in the preface, in the course of a single anatomical demonstration performed comparatively on animals made more than two hundred errors in his description of the parts of the human body and their individual functions.[74]

Theater of Anatomy," *Representations* 17 (1987): 62–95. On Leiden, in particular, see J.-P. Cavaillé, "Un théâtre de la science et de la mort à l'époque baroque: L'amphithéâtre d'anatomie de Leiden," in Working Papers of the European University Institute (San Domenico di Fiesole, 1990). As late as the *Tabulae anatomicae* of Pietro da Cortona, published in 1741 and reprinted in 1788, the title pages of both editions show a skeleton playing an important role in the overall structure of the images.

72. A. Vesalius, *De humani corporis fabrica,* fols. 2r & 3r.

73. Vesalius himself used dogs, pigs, goats, and monkeys for dissection when there were not enough human bodies available to demonstrate the full human anatomy or for performing vivisections in order to show organs in action. In the course of an anatomy held by Vesalius in Bologna in 1540, for example, six dogs were dissected. See B. Heseler, *Andreas Vesalius,* p. 292, and A. Vesalius, *De humani corporis fabrica,* p. 547: "ac postmodum hominem et canem, et, quum obtigit, simiam in omnium praesentia secui, sedulo omnia uti a me describuntur commonstrans."

74. Ibid, fol. 3v: " . . . multo saepius quam ducenties a vera partium humanae harmoniae, usus functionisque descriptione, Galenum declinasse, in unius Anatomes administratione." There are corrections to Galen's anatomy at many points in the *Fabrica,* and Vesalius

(3) The same anti-Galenic polemic can be sensed in the figure of a monkey on the left. The monkey is biting the hand of a bystander, thereby demonstrating its own essential feral qualities. The meaning of this image can be gleaned yet again from the preface, in which Vesalius stresses the fact that Galen had never dissected a human body and had limited himself to animals, especially monkeys. Galen was thus, according to Vesalius, insufficiently aware of the relative morphological and physiological differences between animals and humans, and thereby made a number of serious errors in interpretation and description.[75] This charge is a leitmotiv of Vesalius's anti-Galenic polemic, and the monkey becomes its symbol. Titian is the author of a caricature on the "Laocoon" (fig. 21) considered by H. W. Janson to be part of the controversy between Vesalius and his contemporaries.[76] The woodcut is viewed as a pictorial insult directed against the modern followers of the Galenic tradition. It becomes a *reductio ad absurdum* of Galen's comparative anatomy, taking its cue from a recently discovered anatomically perfect sculpted group of antiquity (1506), which Titian translates into shrieking monkeys.[77] This interpretation of Titian's print has been adopted as further evidence of his direct participation in the realization of the anatomical drawings of the *Fabrica*.

All three features are to be found in the lower part and the foreground of the title page that becomes, according to this reading, an enclosed space given over to the polemic against the ancient practice of anatomy. The positive and innovative section of the image begins immediately above this space, and it is characterized by a more explicit iconographic language. The same purpose can also be drawn from the preface of the

---

repeats frequently that the former had never dissected human bodies. See, for example, pp. 275 and 584, where Vesalius states at length that Galen had never seen the insides of humans, and p. 526, where Vesalius affirms and demonstrates that a large part of Galen's *De anatomicis administrationibus* was incorrect because it was based on the observation of animals.

75. Ibid, fol. 3v: "Ut taceam, mirandum esse maxime, in multiplici infinitaque humani corporis organorum et simiae differentia, nullam nisi in digitis ac poplitis flexu, Galenum animadvertisse." At pp. 42 and 81 Galen is described as "simiarum sectionis quam hominum studiosior," and at p. 83 Vesalius repeats the accusation that the physician of Pergamum and his followers confused human anatomy with that of monkeys. To avoid similar misconceptions, at p. 528 he summarizes the most obvious differences between them.

76. H. W. Janson, "Titian's Laocoon Caricature and the Vesalian-Galenist Controversy," *The Art Bulletin* 28 (1946): 49–56. See also H. W. Janson, *Apes and Ape Lore in the Middle Ages and the Renaissance* (London, 1952), pp. 355 ff. Vesalius's treatise substantially contradicting Galen's works was the object of bitter attacks, reviewed in the final chapter: J. Du Bois, *Vaesani cuiusdam calumniarum in Hippocratis Galenique rem anatomicam depulsio* (Paris, 1552); F. Pozzi (Franciscus Puteus), *Apologia in anatome pro Galeno, contra Andream Vesalium Bruxellensis* (Venice: F. de Portonariis, 1562); G. Cuneo, *Apologiae Francisci Putei pro Galeno in Anatome examen* (Venice: Franciscus de Franciscis, 1564).

77. See also M. Muraro and D. Rosand, *Tiziano, scheda 49*.

Figure 20   Andreas Vesalius, *De humani corporis fabrica libri septem* (Basel: Johannes Oporinus, 1555; Biblioteca Angelica, Rome). For the second, 1555 edition, of the *Fabrica,* also published in Basel by Johannes Oporinus, all the illustrations printed in the first edition from woodblocks have been re-engraved on copper. This provides an occasion to introduce certain modifications in the title page, such as, for example, making the metaphor of the triumph of death more explicit by transforming the magisterial baton held by the skeleton into a scythe.

Fabrica: the recto and verso of its first two sheets are entirely applied to the story of the degeneration of anatomical science and to the presentation of all the polemical texts against this tradition. The balance of the preface, however, provides a proper introduction to the work and its intentions in its dedication to Charles V.

Figure 21   Titian, caricature, "Laocoon" (Museo Civico, Bassano del Grappa). The controversy that toward the mid-sixteenth century saw Vesalius clashing with the large contingent of Renaissance Galenists—indignant about the irreverent criticisms of Galen contained in the *Fabrica*—also involved Titian. The artist knew Vesalius: if Titian himself did not draw and engrave the illustrations of the *Fabrica*, it was at least an artist working or who had worked in Titian's Venetian studio who did so. Vesalius was especially critical of the fact that Galen had seldom observed the human body and that the latter wrote his anatomical works based on his dissection of animals—monkeys in particular. These reproofs naturally also implicated Renaissance Galenists, such as Jacques Dù Bois, for example, who did not dare to question Galenic authority, even though the direct dissection of the human body occasionally showed it to be fallacious. Titian took up the theme of Vesalius's anti-Galenic assault, transforming into an apish caricature the most famous and anatomically perfect sculpture of classical antiquity, the *Laocoön*.

If those features of the title page that I have considered so far seem already quite innovative when contrasted with the iconographic tradition of the anatomy lesson, the scene taking place on the anatomy table represents its overthrow and the transformation of the relationship between theory and practice in the teaching and study of human anatomy. The "ridiculous custom of scholars" that provided for a separation between the space of the spoken word and the space of dissection, and for a differentiation in the hierarchy of the roles around the body, is replaced, for the first time in the history of anatomical iconography, by the image of the teacher dissecting the cadaver. Vesalius is portrayed with his right hand thrust in a woman's abdomen and his left raised as if to emphasize the words accompanying the demonstration. On the table are those surgical instruments necessary for the procedure: a razor and scalpel; next to them, beside a candle, are an inkwell, a pen, and a sheet of paper.[78] The

78. In the title page of the 1555 edition there are a few lines of simulated writing.

same objects appear in the portrait of Vesalius that follows the title page in the first two editions of the *Fabrica* (fig. 19).[79] It shows him at the age of twenty-eight in 1542, demonstrating the anatomy of the forearm (especially the flexing muscles of the fingers) on the body of a disproportionately large woman. Here, too, next to some surgical instruments, is a pen in an inkwell and a sheet of paper on which the words "Of the bending muscles of the fingers. Chap. 30," have been written, recalling the beginning of one of the chapters of the *Fabrica*.[80]

The representations of Vesalius that show him with his hands on the cadaver emphasize the anatomist's activity as both dissector and writer of treatises, and the indissolubly twofold character of anatomy, a product both of practical ability and theoretical elaboration. It therefore proposes a synthesis of the two components of the discipline that had been traditionally separated. Moreover, the method by which Vesalius conceives the preparation of the text in contradistinction to his predecessors is indicated. While in the anatomy lesson conducted according to the quodlibetarian model it is the text that produces the dissection, here it is the dissection that produces the text. This implies a rejection or at least a suspension of judgment of previous authoritative sources, Galen included. The book, which had earlier represented authority, the academy, and anatomical theory is now in the hands of the public in Vesalian iconography: a student to the left of the skeleton and a bearded figure to the right of it have an open volume in front of them that attracts the attention of yet another. A third book is held, but this time closed, by another bearded figure (above, to the right of the image) who points with the same hand toward the dissecting table as if to indicate that attention generally should be focused in that direction, and not only on the book, which, given the demeanor of the figure, looks like an additional piece of cognitive baggage.

Though some have maintained otherwise, it would not appear that

79. It is also contained in the Latin and German editions of the 1543 *Epitome* and in the *Epistola [ad Joachinum Roelants] rationem modumque propinandi radicis Chynae decocti . . . , pertractans et praeter alia quaedam, epistolae cujusdam ad Jacobum Sylvium sententiam recensens, veritatis ac potissimum humanae fabricae studiosis perutilem; quum qui hactenus in illa nimium Galeno creditum sit, facile commonstret. Accessit quoque locuples rerum et verborum in hac ipsa epistola memorabilium index . . . [A Francisco Vesalio, auctoris fratre, editum]*. (Basel: J. Oporinus, 1546).

80. In reality, these words do not correspond to Bk. II, chapter 30 (myology) of the *Fabrica*. They concern, rather, a rough transcription of chapter 43 of this same book. The motto "Ocyus Iucunde et Tuto" inscribed under the date alludes to the physician's mission and derives from an aphorism of Aesculapius transmitted by Celsus. The true source of this image seems rather to be the time, related in the preface, when Vesalius performed a dissection in Paris while he was still a student.

with Vesalius "the anatomical book lost its central place in the anatomical scene."[81] On the contrary, the fact that the book has not disappeared,
but has been displaced and has thereby acquired increased significance
because of a change of use (parallel to the transformation of the dissection
scene), enhances the figure of the physician searching in the viscera of the
cadaver and suggests another aspect of the recasting of the status of the
anatomist proposed by Vesalius.

IMAGES OF DISSECTION IN THE VESALIAN "MANIERA"

The *De humani corporis fabrica* was too voluminous and too expensive
to replace the anatomy manuals in use in the medical faculties. But this
did not prevent it from circulating widely and from becoming a work of
reference for physicians and anatomists on a par with the great texts of
classical medicine. The illustrations of the parts of the human body and
the depictions of the musculature and the skeletons contained in the *Fabrica* quickly became the models for much of the succeeding iconography.[82]
Similarly, the anatomy lecture displayed on the title page becomes the
basis for a series of printed images found in a number of sixteenth-
century treatises that are clearly inspired by the iconographic composition
inaugurated by Vesalius's text. From 1543 onward the illustrated title
pages of anatomical treatises that portray a dissection scene always show
a physician-anatomist who makes the opening incision on the cadaver
with his own hands, while students, colleagues, and the merely curious
surround the table, eager to learn the secrets of nature hidden within the
body.

Juan Valverde de Hamusco, who had studied at Rome with Realdo
Colombo and Bartolomeo Eustachio, is certainly one of the most faithful
followers of Vesalian anatomy. He was so faithful, in fact, that his anatomical treatise, *Historia de la composicion del cuorpo humano*, published in Rome in 1556, can actually be considered to have plagiarized

81. See W. Artelt, "Das Titelbild," p. 69, and L. Premuda, *Storia dell'iconografia*, p. 123.
82. It suffices to think of some of the illustrated anatomical treatises published during
the sixteenth century, which were all, in varying measure, indebted to Vesalian iconography:
T. Geminus, *Compendiosa totius anatomie delineatio. . . .* (London: in off. J. Herfordie,
1545); J. de Valverde de Hamusco, *Historia de la composicion del cuorpo humano* (Rome:
per A. Salamanca and A. Lafreri, 1556); F. Coïter, *Externarum et internarum principalium
humani corporis partium tabulae, atque anatomicae exercitationes, observationesque variae, novis, diversis ec artificiosissimis figuris illustratae* (Nuremberg: T. Gerlach, 1572); F.
Platter, *De corporis humani structura et usu libri III . . . Tabulis explicati iconibus . . .
illustrati* (Basel: Froben, 1581). On plagiarisms of Vesalius, see R. Herrlinger, *History of
Medical Illustration*, especially pp. 121 ff. and L. Choulant, *History and Bibliography*.

the *Fabrica*.[83] A majority of the engravings in Valverde's book, which were drawn by Gaspar Becerra and transferred to copper plates by Nicolas Beatrizet, are more or less copied from it. This is how the anatomist justified his plagiarism in a letter to his readers: "Although some of my friends were of the opinion that I should make new figures, without using those of Vesalius, I chose not to do so, to avoid the confusion that might have followed from it, since then one would not have known so easily in what I agreed or disagreed with him."[84] Valverde thus places his own work fully in the ambit of that of Vesalius, constantly emphasizing the importance of direct anatomical observation on cadavers, the use of illustrations in anatomical treatises, and the performance of dissection by the anatomist.

The title pages of some of the editions of Valverde's *Anatomia* provide further means for evaluating, if this is still necessary, the strict continuity between his text and the new anatomical method introduced by Vesalius. In the engraving at the beginning of the first edition of the Italian translation printed in Rome in 1560 (fig. 22), two skeletons support the elliptical shield that contains bibliographical information.[85] Above it are a monkey and a pig, two animals whose bodies, since antiquity, had often been used in dissections as substitutes for the human cadaver. Bones, skulls, and an hourglass further enrich the image as linking anatomy to the themes of decay and death.

Three scenes are shown in the lower part of the frame. Each illustrates different moments of the anatomy lecture and obviously refers to the pedagogical and investigative principles proposed by Vesalius. In the scene on the left an anatomist is instructing a youth (a student perhaps) on the way to reconstruct a skeleton with the assistance of an illustrated anatomical text. He is basing his instructions on another skeleton that has already been assembled. The book lies on a table with a shelf beneath it that holds some bones, obviously invoking the title page of the *Fabrica*.

83. J. Valverde, *Historia*. Cf. R. Herrlinger, *History of Medical Illustration,* pp. 121–26, and F. Guerra, "Juan de Valverde de Hamusco," *Clio Medica* 2 (1968): 339–63. For a listing of the editions and translations of this work, see L. Choulant, *History and Bibliography,* pp. 205–8.

84. I quote from the Italian translation: J. Valverde, *La anatomia del corpo umano* (Venice: Stamperia de Giunti, 1586), fol. a4r. A. W. Meyer and S. K. Wirt ("The Amuscan Illustration," *Bulletin of the History of Medicine* 15 [1943]: 667–87) have made a careful comparison of the illustrations in the two texts.

85. J. Valverde, *Anatomia del corpo humano . . . con molte figure di rame ed eruditi discorsi in luce mandata* (Rome: A. Salamanca and A. Lafreri, 1560). The translation was made by Anton Tabo and revised by Valverde himself. Apropos this translation, the author wrote in the dedication to Philip of Spain, dated May 20, 1559: "and I would have preferred to have it in Latin, if the effort had not seemed so useless to me since Vesalius had written so much in that language."

Figure 22 Juan Valverde de Hamusco, *Anatomia del corpo umano*. (Rome: A. Salamanca and A. Lafreri, 1560; Biblioteca Angelica, Rome). Animals, skeletons, bones, hourglasses, and skulls are some of the decorative elements recurring in the title pages of sixteenth- and seventeenth-century anatomical treatises. In the Italian edition of Juan de Valverde's *Anatomia*, the architectonic border framing the title displays two sacred bones in the trabeation, the form of which recalls the heraldic shields that often punctuate the decorative element in the architecture of the epoch. The choice of the monkey and the pig overhanging the ellipse enclosing the bibliographical information alludes to the fact that these animals were still being used in the sixteenth century in the public and private anatomy lessons. The dissection of animals frequently took the place of human dissections. On one hand, this was due to the scarcity of human cadavers available for the purpose; on the other, that for animals vivisection was permitted. Three anatomical scenes are shown at the base of the structure: at the left two persons in the act of reconstructing a skeleton, in the center a public anatomy lesson, at the right (detail below), a "private anatomy."

In the larger scene the anatomist surrounded by teachers and students stands at the center and performs a dissection at a public anatomical ceremony. A colleague at the head of the cadaver has an open book before him, a constant feature of the iconography of the anatomy lesson. The vignette on the right shows the physician dissecting the cadaver of a woman in the presence of four more figures. This might be one of those private anatomies whose pedagogical merits Vesalius had praised.[86]

This title page was reprinted frequently in successive editions of Valverde's anatomy, sometimes with slight modifications that, however, leave the general iconographic structure of the image unchanged.[87] A different title page was used for the 1586 reprinting (fig. 23), although some of its features were copied from the 1560 engraving.[88] In the frame, which is structurally similar to the preceding one, two caryatids have replaced the skeletons. The monkey and the pig, above, on the tympanum, now support the outer skin of a cadaver rather than two bones. Below, at the center, between two seated skeletons, other scenes are represented. One on the left shows two persons in an animated discussion over a book open on a table; the one on the right, instead, is a faithful reproduction of a vignette from the 1560 title page. The portrait of Valverde printed on the following leaf of this Giunti edition has the same architectural frame as the 1560 title page, and even the vignette on the right is similar (fig. 24). To this is juxtaposed another on the left, as yet unpublished, which shows a figure kneeling and an instructor who holds an anatomy book before him (note the small human figures that are included to make it recognizable) caught in the act of comparing the text to a woman's body.

In the two vignettes contained in the title page and in the portrait of the 1586 edition of the *Anatomia,* as well as in those of the 1560 engraving, the two common features, the book and the cadaver, found in the traditional iconography of the anatomy lesson are reproduced. Both enhance the various woodcuts that illustrate Mondino's *Anatomia* and crop up again in the images used in Valverde's books, but they are used in another manner, since they appear after the renewal of the discipline

86. For example, at the beginning of Bk. V, chapter 19, Vesalius writes: "enim privatam et inter paucos exhibitam sectionem, publicae praeferendam nemo ambigat." (*De humani corporis fabrica,* p. 547).

87. It is contained: in two editions of the Latin translation by M. Colombo and J. Valverde, *Anatome corporis humani* (Venice: Stamperia de Giunti, 1588–89; 1607); in the Italian edition printed (Venice: appresso Nicolò Pezzana, 1682). The same title page was also used in the Cologne, 1600 edition of Vesalius's *Epitome.* See G. Wolf-Heidegger and A. M. Cetto, *Die anatomische Sektion,* nn. 137–39 and R. Herrlinger, *History of Medical Illustration,* p. 124.

88. J. Valverde, *La anatomia.*

Figure 23   Juan Valverde de Hamusco, *Anatomia del corpo umano* (Venice: Stamperia Giunti, 1586; Biblioteca Angelica, Rome). In the new edition of Valverde's *Anatomia* published in Venice by the Giunti in 1586 the title page is somewhat altered compared to previous versions. On the tympanum between the monkey and the pig the skin of a flayed satyr hangs from the architectonic border. This is an obvious reference to the myth of Apollo and Marsyas recounted in Ovid's *Metamorphoses*, a recurring theme in Renaissance and Baroque anatomical iconography. In the lower part of the frame—between two seated skeletons—two small scenes are being enacted that capture different points in the activity of the anatomist: that of dissection, performed in person, and that of the discussion over texts.

brought about by Vesalius. Valverde's images of dissection therefore incontrovertibly demonstrate, even more obviously than had been the case in the *Fabrica,* the redefinition of those essential and traditional tools for knowing and understanding the composition of the human body.

The new use of the book and of the cadaver are represented in a broadside, dated 1581, preserved in the University Library in Glasgow (fig.

Figure 24   Juan Valverde de Hamusco, *Anatomia del corpo umano* (Venice: Stamperia Giunti, 1586; Biblioteca Angelica, Rome). Juan de Valverde had come to Italy from Spain to study medicine, and in Rome he had been a disciple of Realdo Colombo. In the dedicatory letter of the first edition of his anatomical treatise he mentions that the dissection of the human body was prohibited in Spain. To make knowledge of anatomy more accessible to his countrymen he had written his book in Spanish, rather than in Latin as was customary in educated and academic circles, and published it with a considerable iconographic apparatus. In fact, Valverde records that dissection of the cadaver provoked such uneasiness even for those Spaniards who had come to study the procedure in Italy, where it was permitted, that many of them could not endure it.

25).[89] It shows the English anatomist John Banister, author of *The Historie of Man* published in London in 1578, during an anatomy lesson. He has a cadaver that has been cut open before him, and with his left hand he

89. *John Banister's Anatomical Table,* Glasgow, University Library, Hunterian Manuscripts. Cf. G. Wolf-Heidegger and A. M. Cetto, *Die anatomische Sektion,* n. 250.

Figure 25   *The Anatomical Table of John Banister* (Glasgow: University Library). This scene of dissection is one in a series of paintings on paper commissioned by the anatomist John Banister and executed probably by Nicholas Hilliard in about 1580. Banister portrays himself in 1581 while performing an anatomical demonstration: one of his hands is touching the intestine of the cadaver (note that the names of some anatomical parts are inscribed on the organs), and with his baton he points out the abdominal cavity in a skeleton. On the lectern at Banister's shoulder Realdo Colombo's *De re anatomica libri XV* is open to the chapter dedicated to the viscera. Both Hilliard and Banister wanted to assure themselves of the veracity and legibility of the image and took great pains with the internal coherence of the scene.

touches the intestine, the location of which is also indicated on a skeleton. Behind him Realdo Colombo's *De re anatomica* is open precisely at the chapter devoted to the viscera,[90] as if to stress the unbreakable link between a knowledge acquired from books and empirical observation acquired from teaching. Book and cadaver are similarly represented in Rembrandt's celebrated *The Anatomy Lesson of Dr. Nicolaes Tulp* (1632) (fig. 26), whose dependence on Vesalian anatomy and iconography have been brilliantly analyzed by William Heckscher.[91]

The image at the beginning of Realdo Colombo's *De re anatomica libri XV,* printed in Venice in 1559, a year after the death of the author (fig. 27),[92] is one of the most beautiful and intriguing representations of the

90. Colombo's text exerted an enormous influence on the development of anatomy in England. It was, in all probability, brought across the Channel by John Caius, Colombo's fellow student and later his colleague in Padua.

91. W. Heckscher, *Rembrandt's Anatomy.*

92. R. Colombo, *De re anatomica libri XV* (Venice: Nicolò Bevilacqua, 1559). This first folio edition was followed by three others in 8vo without any illustrated title pages: Paris: I. Foucherii Iunioris, 1562; Paris: A. Wechel, 1572; Frankfurt: I. Wechel, 1590.

Figure 26   Rembrandt van Rijn, "The Anatomy Lesson of Doctor Nicolaes Tulp" (Maurit-shuis, The Hague). The dialogue over the dissected cadaver, versus the book that character-ized the iconography of the anatomy lesson from the very beginning, survived into the seven-teenth century. In 1632 Rembrandt was commissioned to execute his famous *The anatomy lesson of Dr. Nicolaes Tulp*. Tulp, a reader in anatomy in Amsterdam, during an anatomical presentation held before the members of the city's corporation of surgeons, demonstrates to the observers the muscles and tendons of the forearm used in flexing the fingers. With his left hand the anatomist illustrates "from life" the action of these muscles. Evidently the painting makes a clear allusion to the portrait of Vesalius contained in his *Fabrica* and *Epitomes* (fig. 19), where the anatomist is demonstrating the same bodily parts.

anatomy lecture produced after the publication of the *Fabrica*. It shows a lecture conducted in perfect harmony with the iconographic and opera-tive guidelines set by Vesalius: Realdo, at the center with his scalpel, is opening the body of a muscular youth whose body has been shaved, as was customary. Only eleven people, all students and colleagues of the anatomist, are present at the dissection, and they are all involved in vari-ous ways in the operation taking place on the dissecting table. Though the overall morphological structure of the image reproduces an anatomi-cal scene on the model of the title page of the *Fabrica,* certain elements, and especially the cherub in the foreground, seem potentially to suggest another level of interpretation for this title page.

Colombo had studied under Vesalius in Padua and had frequently as-sumed the role of *sector* during the public anatomies.[93] Having temporar-ily substituted for his teacher in the anatomy lessons held in January

93. A. Vesalius, *De humani corporis fabrica*, p. 56.

1543,[94] when Vesalius went to Basel to oversee the publication of the *Fabrica,* Colombo was appointed to the chair of surgery and anatomy in October 1544.[95] Between 1546 and 1548 Colombo taught at Pisa, where Cosimo I had instituted the teaching of anatomy,[96] and he later moved to Rome, where he acted as physician to Paul III and took a readership at the university.[97]

The *De re anatomica,* a work famous in the history of medicine because it contained certain important discoveries concerning the circulation of the blood and the female genital system,[98] was written by Colombo almost as a result of his dissecting activity in the three Italian universities: while opening cadavers, he had felt the necessity of organizing what he had observed in a systematic discourse.[99] Moreover, with his treatise he intended to correct errors committed in the description of the human body and to add to Galen's and Vesalius's texts. In the preface to his work Colombo takes as his point of departure the fact that knowledge of anatomy, as with every science, needs to be acquired by a process of accumulation: "all arts and disciplines are advanced by the additions made by posterity."[100] Although this may seem rather obvious today, this statement justified, on the one hand, Colombo's new treatise of anatomy, and, on the other, juxtaposed it to that of Vesalius, whose work in any case it would be necessary to take into account. His antagonism toward

94. R. Colombo, *De re anatomica,* p. 60.
95. The document assigning the chair to Colombo is preserved in AS Ven., *Senato Terra,* l. 33, fol. 128v. The bibliography concerning him is extremely thin. For biographical information, see E. D. Coppola, "The Discovery of the Pulmonary Circulation: A New Approach," *Bulletin of the History of Medicine* 31 (1957): 47–77; L. Thorndike, *History of Magic and Experimental Science,* 8 vols. (New York, 1959), 5:45, 354; C. Colombero in *Dizionario biografico degli italiani* 27:241. On the anatomical treatise, see K. F. Russell, *The "De re anatomica" of Realdus Columbus* (Melbourne, 1953); R. J. Moes and C. D. O'Malley, "R. C.: 'On Those Things Rarely Found in Anatomy,' *Bulletin of the History of Medicine* 34 (1960): 508–28; and A. Carlino, "L'exception et la règle. À propos du XVe livre du *De re anatomica* de Realdo Colombo," in *Maladies, Médecines, et Sociétés. Approches historiques pour le présent,* ed. F.-O. Touati, 2 vols. (Paris, 1993), 1:170–76.
96. A. Fabroni, *Historiae Academiae Pisanae, 3 vols.* (Pisa, 1792), 2:73, note 1.
97. G. L. Marini, *Degli Archiatri Pontifici,* 2 vols. (Rome, 1784).
98. On the circulation of the blood, see: E. D. Coppola, "The Discovery of the Pulmonary Circulation," and W. Pagel, *William Harvey's Biological Ideas* (Basel and New York, 1966). On the female genital apparatus, see T. Laqueur, "Amor Veneris, vel Dulcedo Appeletur," in *Fragments for a History of the Human Body,* ed. M. Feher, et al. (New York: Zone 5, 1989), pt. 3, 90–131; T. Laqueur, *Making Sex: Body and Gender from the Greeks to Freud* (Cambridge, MA, and London, 1990), especially pp. 96–98, 112–13.
99. "Cum mihi in mentem venit post diuturnos, et pene infinitos, quos in secundis hominum cadaveribus suscepi labores, de universa Anatome primum, deinde vero de anatomica administratione, quae observavi, et cum rei natura consentire experimento didici, scribere" (R. Colombo, *De re anatomica,* letter to the reader).
100. Ibid.

Figure 27   Realdo Colombo, *De re anatomica libri XV* (Venice: Nicolò Bevilaqua, 1559; Biblioteca Angelica, Rome). Realdo Colombo had been a student and collaborator of Vesalius when the latter was teaching at the University of Padua; he took his place as teacher of anatomy and surgery in 1544. Colombo had already begun writing his *De re anatomica* in 1548. It is a treatise, as the author himself states, meant to synthesize and correct what had been written about the anatomy of the human body by Galen as well as by Vesalius. The book was published posthumously in Venice in 1559. The title page of the *De re anatomica* is a sort of allegorical regret for a work that came to press without an iconographic apparatus. Michelangelo is portrayed at the right of the image. The cherub is taking his hand, inviting him to replace the artist crouching in the foreground. In fact it was Michelangelo who was supposed to have executed the anatomical figures for Colombo's book.

Vesalius, which is somewhat diluted in the rhetoric of the letter to the reader in the *De re anatomica,* comes out vigorously and explicitly in the body of the work itself, where he belabors the errors committed in Vesalius's *Fabrica.*[101] In a letter he writes to Cosimo I de' Medici on April 17,

101. In Bk. V, chapter 20, Colombo writes: "Vessalius . . . , qui Gal. reprehendit, reprehensione non vacat." Other corrections can be found, for example, in the *De re anatomica*

1548, in which he justifies his absence from the Pisan *Studium,* he states that he had begun "to write a work by which they [those who want to learn about the composition of the human body] will have the truth about the matter, and that it is possible to study anatomy while performing it; seeing the harm that is suffered in studying Galen, who besides being long winded, is mendacious. Similarly, Vesalius is prolix, and not rarely needs emendation, as I have demonstrated in public, and as I am even more openly about to make evident, if Your Excellency does not prevent it." The book in question was naturally the *De re anatomica.*[102]

Of course, Cosimo might have objected; the work could just as well have been written in Pisa. In this way, Colombo justified his continuing Roman sojourn, which he would make permanent a few months later: living in the papal city was indispensable, "both because good fortune made available to me the leading painter in the world, and because of the large number of bodies that it is necessary to have almost continually at hand, so that the subject can be treated properly; and this is because I am having to contradict all, both ancients and moderns."[103] The artist alluded to was Michelangelo, who was supposed to draw the images for Colombo's treatise. Cardinal Ridolfi was the intended patron who would help to see it through the press.[104] After the publication of the *Fabrica,* in fact, it had become impossible to think of producing a useful and successful work of anatomy without a conspicuous illustrative apparatus to accompany the text. Nevertheless, as is well known, the *De re anatomica* appeared a full eleven years after this letter and with only a single illustration, the title page.[105]

Some details of this title page suggest a close connection between the vicissitudes of the illustrations of the *De re anatomica,* and the scene of

---

at pp. 48, 54–55, 101–2, 104, 132, 221, 235, 254–55, 260, 300, 319, 348–49, 405, 442, 447. Cf. R. De Maio, *Michelangelo e la Controriforma* (Bari and Rome, 1978), p. 443.

102. The letter is preserved in the Archivio di Stato, Florence, *Carteggio universale di Cosimo,* fasc. 386, fol. 258. It was discovered and transcribed by A. Parronchi, "Michelangelo e Realdo Colombo," in *Opere giovanili di Michelangelo* (Florence, 1968–92), 2:193–94.

103. Ibid.

104. Both Buonarroti and the cardinal were patients of Realdo. Ridolfi was especially interested in the study of anatomy and medicine. He, in fact, sponsored the Latin translation by Guido Guidi (Vidus Vidi) of the Greek medical works that he owned. On the proposal to publish the anatomical treatise with Michelangelo's illustrations, see: A. Parronchi, "Michelangelo," pp. 195–96; D. Summers, *Michelangelo and the Language of Art* (Princeton, 1981), especially chapters 13 and 15; and J. B. Schultz, *Art and Anatomy in Renaissance Italy* (Ann Arbor, 1982), pp. 102 ff.

105. The title page is certainly not by Michelangelo, as was demonstrated by L. Steinberg, "Michelangelo and the Doctors," *Bulletin of the History of Medicine* 56 (1982): 543–53.

dissection. The first detail to be noted is the figure on the left, who is consulting an open book.[106] He is represented very differently from what we have seen before: we can see that he holds the simulated writing, and also the image of a man with a raised right arm. The artist's intention seems to be to identify the image as that of an illustrated treatise on anatomy. I suggest that it is Vesalius's *De humani corporis fabrica*.[107] In it, four of the *écorchés* assume the same pose as that of the figure here. This alludes to a twofold intention: on the one hand Realdo in performing the dissection is correcting Vesalius's errors; on the other, the *Fabrica* itself, a text enriched by a copious illustrative apparatus such "that it may be studied while performing anatomy," is the typographical model that Colombo is hoping to imitate.

Below, to the left of the title page, another figure, seated on the ground is shown drawing: it is an artist, present at the dissection, who has been entrusted with preparing sketches "from life," to illustrate an anatomical text.[108] In the center foreground the cherub, taking the hand of the person on the extreme right, seems to be inviting him to some act. The latter strikes an attitude that may be one of surprise, modesty, and timidity. This figure probably represents Michelangelo Buonarroti: there is clear resemblance to the numerous portraits of the artist. This would not be the first time that Michelangelo would have been portrayed next to a dissecting table; one need only recall the anatomy lesson attributed by Roberto Longhi to Bartolomeo Passarotti (fig. 28).[109] Michelangelo's passion for anatomy was well-known and is affirmed by his works as well as by his biographers, Ascanio Condivi and Giorgio Vasari, who record how much he would have liked to prepare a treatise on hu-

106. Another book is being consulted by a student at the right of the anatomist.
107. Vesalius's text is recognizable in other representations within the ambit of medical iconography. It can be found in the portrait of Felix Platter (executed by Abel Stimmer and dated 1578) contained in Platter's anatomy treatise and inspired, *ex professo*, both in the text and in the illustrations by the *Fabrica* (*De corporis humani structura*, fig. 29).
In the painting entitled "The Physician" (c. 1653) by Gerard Dou preserved at the Kunsthistorisches Museum, Vienna, Vesalius's book is recognizable by the unmistakable skeleton leaning on a shovel (fig. 30).
108. It has been pointed out to me that the position of the alleged artist is somewhat unsuited for drawing the cadaver lying on the anatomical table from life, and that, really, the figure is of a student taking notes. I would exclude this interpretation for three reasons: first of all, the position of the pen on the paper makes it quite improbable that the student is writing and would imply writing from right to left (unless we want to suppose this an Arabic or Hebrew student!); second, the support on which he is writing does not appear to me to be a notebook, but rather a tablet; finally, it is not unusual to come upon representations of the artist in the act of copying from such "infelicitous" positions.
109. The painting is housed in the Galleria Borghese, Rome.

Figure 28   Bartolomeo Passarotti, "Lezione d'anatomia" (Galleria Borghese, Rome; Archivio Scala, Florence). Michelangelo's interest in anatomy, and the many dissections at which he was present, are discussed at length by Giorgio Vasari (in the second edition of his *Vite*) as well as by Michelangelo's disciple and biographer Ascanio Condivi. From the time of the founding of the *Accademia del Disegno* in Florence (1564), established to train young artists (with Michelangelo himself as a model), human anatomy became one of the disciplines deemed indispensable in the education of the artist. It was taught using skeletons and *écorchés*, but also through dissections. This painting attributed to Bartolomeo Passarotti portrays just such an anatomy lesson for artists.

man anatomy for artists that would complement the work of Albrecht Dürer.[110]

All these elements combine to suggest an interpretation of the title page that goes well beyond the notion of a simple representation of a dissection scene. The cherub, used here to evoke an allegory and as a bearer of sym-

110. A. Condivi, *Vita di Michelangelo Buonarroti, pittore, scultore architetto e gentilhuomo fiorentino publicata mentre viveva dal suo scolare Ascanio Condivi* (Florence, 1746), pp. 9, 45–6, 48–50 (1st ed. Rome: Antonio Blado, 1553), and G. Vasari, *Le vite de' più eccellenti pittori, scultori e architettori*, a cura di G. Milanesi (1906; reprint, Florence, 1981), pp. 268–69, 274. The work dates from 1568. Dürer's work is the "Theory of Proportion" published posthumously in German with the title *Hierinn sind begriffen vier Bücher von menschlicher Proportion* (Nuremberg, 1528). Condivi, confirming the ties between Realdo and Michelangelo and their mutual project, recalls that the anatomist sent the artist "the dead body of a young and handsome Moor. . . . Michelangelo showed me many rare and secret things on this body that were perhaps never heard of again, all of which I have noted" (p. 50). For additional details, see D. Summers, *Michelangelo*, pp. 397–405 and A. Carlino, "The Book, the Body, the Scalpel," especially pp. 46–50. When my article was published, I was not aware of an article in which R. P. Ciardi had also recognized the portrait of Michelangelo in Realdo's book and discussed the title page (R. P. Ciardi, "Michelangelo, come Galeno: Un'ipotesi iconologica," in *Studi in onore di Giulio Carlo Argan*, ed. M. Bonicatti et al. (Rome, 1984), 1:173–81.

Figure 29   Felix Platter, *De corporis humani structura et usu* . . . (Basel: Johann Froben, 1583). Like many other anatomists of the second half of the sixteenth century, Felix Platter aligned himself with the supporters of the anatomical revisionism championed by Vesalius against the Galenists. In a portrait executed in 1578 by Abel Stimmer published in the 1583 edition of Platter's book on human anatomic physiology, the author has depicted himself holding a book by Vesalius.

bols, holds four pens in his left hand. If the person whose hand he is taking is indeed Michelangelo, then the cherub is inviting him with this gesture to take the place of the seated artist and draw what should have become the illustrations for the *De re anatomica*. The image could thereby be read as a sort of allegory of regret for the failed project.

Going beyond the identification of Michelangelo, it is certainly clear that the title page of Realdo Colombo's treatise sought to emphasize vigorously the need to include illustrations in anatomical texts. Anatomy books are destined to be corrected, emended, enriched by an examination of the body. Colombo makes clear in his letter to the reader his belief that anatomy, along with every other science, should continually perfect itself by a progressive accumulation of knowledge so that every treatise must

Figure 30   Gerard Dou, "The Physician"(Kunsthistorisches Museum, Vienna). In 1653 Gerard Dou, a student of Rembrandt's, produced a painting entitled *The physician*. The subject is portrayed analyzing a bottle containing the patient's urine. On the windowsill, prominently displayed, sits an open anatomy text. Dou wanted it to be recognized: the figure of the skeleton is unmistakably one of the images that decorates Vesalius's *Fabrica*.

inevitably be rewritten. This entails not only questioning ancient authorities, as had been done by Vesalius and to some extent by Berengario, but displaying an awareness of the vulnerability and relative importance of one's own work and findings.

This survey of fifteenth-century anatomical images points to a radical shift in the representation of the public anatomy lecture initiated by the publication of Vesalius's work. Despite the change in iconography, which corresponded to the epistemological renewal of this discipline, there was no corresponding revision of the academic ceremony: in spite of Vesalius the anatomy lesson continued to be regulated by university statutes for

several more decades and to be carried out according to the quodlibetarian model that imposed a precise separation in the roles of the *sector*
and *lector,* and a specific distance between the reading of the text and the
action of dissection. It was an iconography not reflected in reality but
primarily based on the image offered by the most important and successful book published in this period. The redefinition, proposed by Vesalius,
of the uses of the cadaver and of the book would therefore not find an
immediate or even an ensuing response in European universities. Obstacles of an institutional as well as an intellectual and cultural nature produced a scission between what the anatomist could and would have
wanted to do.

CHAPTER TWO

Practices: Norms and Behaviors at the Public Anatomy Lesson in the *Studium Urbis*

Along with books, dissection provides the essential apparatus for the teaching and the study of the parts of the human body. The opening of the cadaver for scientific purposes, however, was an operation that had both religious and anthropological consequences. Dissection, where it was practiced, was closely controlled by a series of detailed regulations established by university statutes that were similar all over Europe. The practicability of dissection and the progress of anatomy were thus tied to norms that, in certain respects, openly contradicted the methodological principles enunciated by many anatomists. The regulations shaped the organization of the anatomy lesson, as well as the legal procedures by which the universities procured cadavers and the access to them by physicians and students.

The protagonists of the lesson are the anatomist and the cadaver. Historians of science have dedicated volumes to the study of the medical professions, to the biographies of the great anatomists and, more recently, to the university institutions, but little or no space has been given to the history of the dissected bodies, to an analysis of the statutes regulating them, and to the criteria by which they were selected. In this chapter I propose to treat in the greatest detail possible the ways in which physicians, universities, and religious and political authorities managed and organized the practice of dissection. I will pay special attention to those aspects and procedures that pertain to the cadaver during and after the operation, and will provide a short history of anatomy that also takes into consideration the point of view of the cadaver, which has a physiognomy and a name.

For this purpose it is necessary to choose a specific and significant but

69

limited context. Rome is ideal for at least two reasons. First of all, the *Studium Urbis* in the sixteenth century was one of the most important universities for the teaching of anatomy, numbering among its teachers such authorities as Realdo Colombo and Bartolomeo Eustachio, who have been consecrated by the historiography of science. Second, the papal city can serve as an ideal vantage point from which to examine all those religiously ambiguous factors connected to the practice of dissection and the handling of the deceased, and to identify the strategies, institutional or not, used to circumvent them.

BETWEEN THE CURIA AND THE COLLEGE:
A PORTRAIT OF THE PHYSICIAN

For a *lector* the teaching of anatomy, and thus the possibility of dissecting cadavers, usually assumed a secondary role as compared to the exercise of medicine in the broader and more pertinent sense of the term and as compared to the teaching of other subjects such as theoretical medicine, practical medicine, or surgery. For many years the anatomist remained hidden behind the more general figure of the physician, just as anatomy was considered an auxiliary discipline that did not require particular specialization and certainly did not merit having a chair specifically assigned to it.

Any attempt to define the role of the anatomist within the structure of the university and of society, therefore, must be preceded by a thorough understanding of the status of the physician.

The *Collegium Medicorum Almae Urbis* was the institution that controlled and regulated the activities of physicians in Rome. The year of the founding of the college is not known,[1] but the first official documents that specifically refer to it, and assign responsibility for administering the activity of physicians in professional, juridical, moral, and corporate terms, date to the pontificate of Sixtus IV (1471–84).[2]

The *Collegium Medicorum* was composed of twelve persons, of whom eight senior physicians, *antiquiores doctores,* were called *numerari,* since they divided among themselves money received from those taking doctorates. Of the remaining four, two others were also *numerari,* and the rest were *supernumerari.* Each year, on January 1, the oldest of the *antiqui-*

---

1. In the letter of dedication to Clement X published in the 1676 version of the *Statuta,* it is stated, apropos the foundation of the college, that it had been "a tempore immemorabili constitutum" (fol. 4v). In the first chapter of these statutes the origins of the college are said to date to the days of imperial Rome (p. 3).

2. *Statuta 1676,* fol. 4v.

ores doctores was elected *protomedicus,* or "Prior and Senior Physician General of all the Cities of the Church."[3] He was without doubt the official possessing the greatest power in the field of medical practice.[4] "He held first place among the rest of his colleagues and the other physicians";[5] the college itself could not function in any way in his absence.[6] At the end of his term his successor was appointed on the basis of age seniority. He was supported by two *consiliarii*—his predecessor and successor in the position of *protomedicus*—and certain decisions could not be made without consultation or even actual agreement among them (as occurred, for example, in the granting of permanent licenses for the practice of medicine).

In the course of the sixteenth century, with the growth of the curriculum and the parallel increase in the student population in the *Gymnasium Romanum,* the college became the only organism for selection and control. It performed its duties rigorously and thoroughly in the assignment of the degrees of *doctor* or *magister* and *licentiatus.* The college itself subjected anyone who, after university study, intended to practice medicine in territories under its jurisdiction to an examination. The entire process is described in detail in the statutes contained in the *Bulla de Protomedici et Collegii Medicorum Urbis iurisdictione et facultatibus* promulgated by Clement VII (1523–34) in 1531, which remained in force at least until 1676, when the statutes were republished with the necessary additions and revisions and were "more solidly consolidated" by Clement X (1670–76).[7] The regulations applied to those who planned to become

3. In the *Statuta 1676* (chapter 8, p. 19) the *protomedicus* was to be included among the *supernumerari.*

4. "Eius [protomedicus] autem iurisdictioni non solum Collegae, verum etiam (quoad ea quae ad Artem medicum quacumque ratione referentur, aut pertinent) non solum Urbis, sed omnium quoque Civitatum, et Terrarum tam mediate, quam immediate S. R. E. subiectarum Medici, et quicunque aliqua ratione medicinam, vel eius aliquam partem exercent, et ipsi quoque Aromatarii tam Urbis, quam totius Status Ecclesiastici, et quicunque aliquid humanis corporibus valetudinis gratia administrant, subijciuntur (*Statuta 1676,* chapter 8).

5. With the exception of the person who held the position of papal physician (ibid). But we shall see that frequently the same person filled the position of *protomedicus* and papal physician contemporaneously.

6. "Quod in Collegio nihil possit fieri sine Protomedico," excepting when the *protomedicus* had to be absent for a considerable period of time and appointed a *vice-protomedicus* to fulfill his responsibilities. See *Statuta 1531,* chapter 4.

7. This bull, containing the 1531 Statutes, is preserved in the library of the ASR among the collection of statutes and in the section *Università,* b. 58. Both are in a 1627 printed edition (Romae, Ex Typ. Rev. Cam. Apost.). The same library also has the manuscript *Statuta Almae Urbis Medicorum* (Collection of *Statuti,* n. 369/2). The 1676 statutes harken back to those of 1531 without significant differences, as we deduce from the letter of dedication to Clement X contained in the same booklet (fols. 4v–5r).

doctors of medicine and surgery as well as of philosophy, "jointly or separately," as usually occurred in almost all universities in the sixteenth century. Only the *Archiatrorum Romanum Collegium* (The Roman College of Physicians) had the authority to confer the *dignitas doctoris*.[8] The candidate who aspired to a degree in philosophy and/or medicine "should show that he had dedicated himself to philosophy" and, for medicine, should have studied for at least two years at a public university, proving this to the *protomedicus* and to the college with documents attesting to his attendance at the courses. Depending on the status to which the candidate aspired (philosophy and/or medicine, surgery) he had to go through an oral exam, consisting essentially of the discussion of certain texts, that was conducted at the church of Sant'Eustachio.[9] The statutes of 1676 also imposed a practical exercise, of which there was no mention in 1531: after the textual phase was over, one member of the college put "a case of illness so that he may furnish a prognosis and course of healing" to the candidate. Then the college, behind closed doors and by secret vote, evaluated the candidate's suitability. Finally, after the conferral of the degree and the laying on of the *insignia,* the new *Magister* had to take an oath of absolute compliance to the college and to the conventions sanctioned by the statutes.

To avoid the usurpation of the "title of physician" in the practice of medicine by fraudulent doctors who placed at risk the health if not actually the life of indigent patients who might put themselves in their care, Sixtus IV, at the explicit request of the college, decreed with a brief dated January 19, 1471, that no one would be permitted to exercise the medical arts in papal territory without first passing an exam attesting to his fitness.[10] This regulation remained in effect throughout the sixteenth and

8. " . . . et nullus in Urbe Roma habeat potestatem doctorandi aliquem, nec possit, nisi interveniant Protomedicus et Consiliarii" (*Statuta 1531,* chapter 27). In this regard, see also the constitution of Julius III dated April 21, 1583, which granted the college a monopoly over degrees in philosophy and medicine. No other authority could claim any competence in the conferral of the doctoral rank (*facultas doctorandi*).

9. For both medicine and philosophy the examination was based on the first and second parts of the *Aphorisms* of Hippocrates and on the first and second books of Aristotle's *Physics;* for the *gradus* in medicine, the *Aphorisms* and Galen's *Ars Medica* (*Tegni*); for philosophy, the works of Aristotle and the *Analytica Posteriora;* for surgery, once again the *Aphorisms* and the *Ars medica,* but only those parts that concerned the discipline (*Statuta 1676,* chapter 34; cf. *Statuta 1531,* chapter 27).

10. "Nemo sive masculus, aut foemina, sive Christianus, vel Iudaeus, nisi Magister vel licentiatus in Medicina foret, vel saltem a Priore dicti Collegii Generali Protomedico, eiusque Consiliariis examinatus, et approbatus existeret, auderet humano corpori mederi in Physica, vel in Chyrurgia in Terris, et dominiis eiusdem Sanctae Romanae Ecclesiae" (Brief of Sixtus IV, "Variis quamquam distracti curis . . . ," January 19, 1471). A number of copies are preserved in ASR, *Università,* b. 25.

seventeenth centuries. Moreover, the 1531 statutes determined that whoever desired to obtain a license to practice medicine, surgery, or even serve only as a barber, had to be examined by the *protomedicus* (joined on occasion also by the two counselors). If the candidate was considered suitable he was issued a written license valid for three years, at the end of which he had to present himself once again and obtain an extension. Occasionally licenses of life duration were granted, but only in cases in which the ability of the physician in the exercise of his profession, and his moral conduct, were absolutely beyond reproach and certifiable as such.[11] By these conventions the college retained absolute control over the practice of "physical" medicine and surgery in the papal state.

The statutes issued a series of requirements for the admission of new members to the college, which the aspiring "colleague" had to fulfill. Most of them were concerned with the moral rectitude of the physician and of his family, his physical health, and conduct that was of a high juridical, religious, and professional order.[12] The physician had to be at least thirty years of age, have an annual income above one hundred golden *écus*, give evidence that he had a permanent residence in Rome and "that he had the customary documents, or permission to teach if he had in the previous decade achieved the status of Doctor in any well-known university."[13] After the qualities of the candidate had been ascertained, he was subjected to an examination in which he had to discuss and interpret a number of *conclusiones* in medicine and philosophy.[14]

The rubric in the statutes of 1531 dedicated to the admission of new members into the college, specified, instead, that: "No one who is not a Roman born of Roman parents may belong to the College, except for what is otherwise specified in the Statute concerning candidates for the

11. See Clement VII, *Bulla*, chapter 57.
12. "Conditiones autem sunt probitas in primis, et morum integritas; adsit eruditio non vulgaris; . . . defectu natalium non laboret, nec parentibus erubescat sordidioribus, vel utrique, vel altero; nec aliquo mutilatus membro, aut depravato ridiculus videatur; nulla unquam fuerit haeresi, aut aliqua iuris, vel facti infamia notatus; nec alteri in ludo litterario inservierit; nec sordidam artem aliquando exercuerit; nunquam studiose fecerit homicidium; nunquam contra Romanum Medicorum Collegium, aut contra iura eius molitus sit, aut moliri tentaverit . . . ; tam viventis Collegae, quam actualiter Exercentis Aromatariam filius, aut filii filius, aut frater, aut nepos non sit . . . (*Statuta 1676*, chapter 19).
13. Ibid.
14. "Collega novus, quo die admittendus est (congregato ex auctoritate Protomedici Collegio) de loco primum eminenti aliquod Medicinae, vel Philosophiae argumentum suo arbitratu, interpretetur. Duas quoque conclusiones defendat, alteram ex Medicina, alteram ex naturali Philosophia, quas Protomedicus spacio ante congruo ad duos ex Collegis mittet, qui de eis cum Collega admittendo duabus saltem argumentationibus propositis disputabunt" (ibid, chapter 21).

College. However, the frivolous, the disreputable, Jews, those without a doctorate, bastards, frauds, and perpetrators of incest will be considered ineligible, so that the Roman College and *Studium* may be preserved, and all disrepute be perpetually avoided." [15]

When the 1531 statutes are compared with the corresponding section from 1676, the conditions for admission set forth in the earlier statutes actually seem exclusionist. The 1531 statutes were more preoccupied with the circumstances of the candidate's birth than with his professional qualities, and there is no mention of an examination. Behind all this and especially in the clause concerning citizenship there is a clear intention to establish protective regulatory control over the profession and perhaps also over local technical traditions, somewhat along the lines of the conventions that regulated contemporary occupational guilds.

The restriction that only Romans could be admitted to the college disappears in the 1676 statutes. Moreover, a list of members demonstrates, to cite but one example of a non-Roman who was granted membership, that as early as 1533 a Genoese, Scipione De Manfredis, had been elected as *protomedicus*.[16] This means that he had entered the college at least by 1530, since to be elected senior physician one had to have been a member for at least three years. This would suggest that the rule had never been consistently applied, probably because there were famous medical men who had come to Rome at the invitation of the pope to serve as his personal physician or as university professors who were offered a place in the college.

At this point it is important to define the relationship that existed between the college and the university. Examining the prerogatives the college had over the *facultas doctorandi* and the selection of students, it would seem that the members of the college who constituted the commission overseeing the conferral of university degrees in medicine, surgery, and philosophy would also occupy the university chairs in medicine. But nothing is said in the statutes concerning teaching duties and the recruitment of the readers of the *Studium*. Moreover, none of the chapters mention the regular enlisting of teachers for the college, nor, for that matter, did membership of the college necessarily impose teaching duties. There was no juridically explicit identification of the college with the teaching body of the *Sapienza,* at least for the period in question. However, it suffices to compare the list of physicians of the *collegium* with that of the

---

15. *Statuta 1531,* chapter 51.
16. "Nomina et cognomina DD. Collegii Romani Medicorum quorum memoria extat," in *Statuta 1676,* pp. 105 ff.

lectores, to observe that the same names are frequently repeated in both, further proof of the close connections between the two institutions.[17] In addition, if we leaf through the minutes of the college's *Liber decretorum,* it is clear that some of the lecturers were present at its meetings.[18]

The regulation limiting affiliation with the college to Romans would seem to have represented an obstacle to the election of a teacher at the *Sapienza* as a member, since those teachers were rarely citizens of the papal state. If this rule had been effectively applied it would have provided a neat separation between the instructional function (which belonged to the *lectores,* usually foreigners) and that of examining and conferring degrees (the competence of the *collegium,* composed of Romans). In fact, leaving aside the regulations contained in the statutes, there was a crossover of duties whenever the *lectores* assumed the role of examiners, although the responsibilities of the *lectores* and those of the *collegium* remained distinct. At this point the clause requiring Roman citizenship was not enforced.

Evidence of this is provided by the statutes of the University of Padua. Chapter 25 of book II establishes that attendance is required "not only at the beginning of the academic year, but also in all disputations, repeti-

17. Both inventories are extremely fragmentary and reflect the state of documentation on this subject, so I have decided not to publish them. I have reconstructed the list of *collegii* from the entries under the *Nomina et Cognomina* in the 1676 *Statutes,* pp. 105 ff. The list of *lectores* in medicine (theoretical medicine, practical medicine, surgery, anatomy, and temporary lectureships (*letture straordinarie*) in sixteenth-century Rome is based essentially on the names furnished by I. Middendorpius, *Academicarum Orbis Christiani* (n.p., 1583); G. Carafa, *De Gymnasio Romano et de eius Professoribus . . . libri duo* (Rome, 1751); F. M. Renazzi, *Storia dell'Università degli Studi di Roma, detta comunemente la Sapienza* (Rome, 1804). Biographical data has also been obtained from A. Portal, *Histoire de l'anatomie et de la chirurgie,* 6 vols. (Paris, 1770–73); A. von Haller, *Bibliotheca anatomica, qua scripta ad anatomen et physiologiam facientia a rerum initiis recensentur* (Zürich, 1774–77; reprint, Hildesheim and New York, 1969); J. Douglas, *Bibliographiae anatomicae specimen, sive catalogus omnium pene auctorum qui ab Hippocrate ad Harveum rem anatomicam . . . scriptis illustrarunt* (London, 1715). Through archival digging I have succeeded in correcting and adding to both lists. For the readers (*lettori*) I have consulted, in particular, the rosters (*Rotoli dei Ruoli*) housed in the ASR, *Università,* "Cimeli" and the reports of the head beadle housed in ASV, Arm. XI, t. 93a. For the members of the College, see ASR, *Università,* "Libri decretorum." The following physicians, who were also members of the *Collegium,* taught in the *Studium* in the sixteenth century: Matteo Corti, Giovan Battista Veroli, Sebastiano Veterano, Alessandro Spinosa, Severino Sillano, Giovanni Antracino, Francesco Leopardi, Giustino Finetti, Bartolomeo Eustachio, Girolamo Cardano, Guglielmo Giscaferri, Francesco Ginnasio, Ippolito Salviani, Pietro Crispo, Virgilio Riccardi, Sallustio Salviani, Ferrante Eustachio, Giacomo Lampugnani, Andrea Cesalpino.

18. An example: on December 28, 1566, Stefano Cerasio, the departing *protomedicus,* hosted a dinner to celebrate the transfer of the position to his successor, Panunzio Sillano. Eight other members were in attendance. Two of them, Ippolito Salviani and Francesco Ginnasio, were *lettori* at the *Sapienza.* See ASR, *Università,* b. 48, fol. 2r.

tions and public activities [of the university] not only by instructional staff (*doctores legentes*), but also by the prior of the College and all its other components,"[19] emphasizing once again the required presence of both institutions at official and public events. It does not seem unreasonable to suggest that the same type of relationship was also in effect at the University of Rome.

It is difficult therefore to assess the relationship between college and *Studium*. From the records, at least, it appears that the first is part of the second. The college possessed the *facultas doctorandi* and controlled medical practice over the entire papal state, while the *Studium* controlled the education of future physicians. However, it was the college that fixed the place, time, and modalities of the public anatomies which, logically, should have been included in the instructional domain and remained within the jurisdiction of the university.[20] It is as if both college and university, which were created to fulfill specific functions in complementary fields (instruction/examination, education/professional practice) had eroded their distinct roles by overlapping their activities, using the same people for both teaching and certification, whether for the education of the physician or for the control of his occupation, perhaps to provide more homogeneous procedures, directives, and intentions. As a consequence the two institutions lost their original features along the way.

The problem of the recruitment of the *lectores* remains open, and the Roman documentation is particularly thin in this area. The formula usually followed for the conferral of a university chair was that of an appointment. The pontiff himself sometimes expressed his own preferences for this or that physician, frequently on the basis of recommendations made by other *lectores,* the college, or the *bidello puntatore,* head beadle, the chief administrative officer of the *Sapienza*.[21] Such nominations were frequently preceded by a petition on the part of the aspiring instructor, by a rescript from the authorities (pope, principal of the *Studium*), and by negotiation over the salary. *Lectores,* in addition to their merits or the connections that had determined their election, had to possess, like the other members of the college, the soundest moral reputation and to be known for their religious observance.

Another category of medical personnel that we need to consider, if only

19. *Stat. Gymn. Pat.,* chapter 25. They are divided into four books: the first three are the 1465 statutes, the fourth contains subsequent corrections.
20. See the succeeding paragraph.
21. The head beadle prepared a report for the pope annually in which he commented on the educational situation and the conduct of students and instructors during the course of the academic year. Occasionally he also wrote to the pope suggesting replacement of teachers who were not fulfilling or were unqualified for their duties, and proposed substitutions on the basis of data they had themselves collected.

because of the preeminent position of those who fell under it, is that of the papal physician. The papal physician was usually someone of established professional reputation in the service of a prominent member of the papal court, or perhaps of the pope himself before his election, or of one of his predecessors. These physicians became familiars of the pontiff. The position of papal physician was normally accompanied by the conferral of rich benefices, or other prestigious duties, and was certainly well remunerated. The papal physician was a member of the college and was also frequently associated with the university. As soon as someone was appointed as physician to the pope, it was arranged for him to receive a chair in the *Studium*.[22]

In the sixteenth and seventeenth centuries in the papal state the activity of physicians, and consequently also of anatomists, was controlled by a select group of *doctores* (the *protomedicus*, members of the college, *lectores*, and the papal physician) through a pervasive normative and institutional organization. The general regulations that provided for this control and the more important decisions, such as, for example, the hiring of new *lectores* for the *Studium*, were instead directly sanctioned by the pope. This pyramidical structure that had at its summit the head of the church, even if at times only as a formality, imposed an ethical and professional code of conduct on physicians, as the statutes demonstrate, in line with the religious principles on which the entire machine of the papal state rested. Any act on their part, whether diagnostic, prognostic, therapeutic, didactic, or involving research, could not contravene these statutes. Such acts inevitably included the practice of dissection.

PRELIMINARY PROCEDURES AND PUBLIC CONTROL

Chapter 18 of the 1531 statutes asserts that "no one in Rome can perform an anatomy without the permission of the *protomedicus* and of his counselors."[23] However, chapter 64 of the very same statutes reveals that it is the *protomedicus* who "ordered the anatomy" whenever he consid-

---

22. For a listing of the activities and roles of papal physicians, see P. Mandosius, *Theatron in quo Maximorum Christiani Orbis Pontificum Archiatros . . . spectandos exhibet* (Rome, 1696), and G. L. Marini, *Degli Archiatri Pontifici*, 2 vols. (Rome, 1784), vol. 1.

23. He continues: " . . . quod si secus fieret a quocumque toties quoties contrafactum fuerit, decem ducatorum Collegio applicandorum poena mulctetur" (*Statuta 1531*, chapter 18). It is interesting to note that anatomists specifically appointed by the pope were exempted from this fine ("exceptis Professoribus in hoc Collegio ad hoc specialiter conductis"). See the decree of August 9, 1571, in ASR, *Università*, b. 48, fol. 76, and *Liber novorum reformationum seu statutorum . . . 1582* (henceforth *Statuta 1582*), MS in ASR, *Università*, b. 61, pt. 3, chapter 54. In this revision of the statutes the entire college, together with the *protomedicus*, granted the licenses for the performance of the anatomies.

ered it appropriate. It had to be performed publicly "at a suitable time" and "in a place where it could best be observed." In addition to medical students, attendance was permitted to "all surgeons and barbers carrying out phlebotomies." They were required to pay a "tax," which was basically an admission fee to a spectacle. Because of the didactic function of the demonstration, the presence of barbers was not only allowed but was actually required in order to prevent that their ignorance of the organization of the human body might result later in an operation on a patient that did more harm than good.[24] The funds collected from the admission fees were used to cover the expenses of the dissection and "if there is money to spare let it be given to the poor of Christ for the soul of the anatomized."[25]

The procedure to be followed was laid down, at least until the statutes of 1584,[26] in eight sections:

(1) "A license should be obtained from the Vicar of the Pope";
(2) "The body for dissection is to be obtained from the Senator or Governor";
(3) "The various categories (*fiant Capitula*) must be listed lest it [the dissection] be held confusedly, in accordance with the customs and suitability of the times";
(4) "The place must be prepared, with instruments, benches, tables . . . and similar necessary things";
(5) "The person who is to be the *lector* should be chosen, and should be either one person reading continuously, or several each in place, as may seem best to the Doctors of the College";
(6) "Two *censors* should be chosen, one of whom should serve as demonstrator, the other as dissector";
(7) "Someone should be available to transport the pieces to be buried";
(8) "At the end of the anatomy funeral rites should be held and at least twenty Masses celebrated for the soul of the deceased."[27]

The organization of the public anatomy lecture thus not only involved the academic authorities and those of the college but also representatives

24. In a note left by Lorenzo Garloni (?), who was *protomedicus* in 1619, we read: "Nelli statuti del collegio è espressa autorità di fare l'anatomia ogn'anno e però bisognandovi spesso la autorità di gravar Barbieri e Chirurgi a contribuire alla spesa e ritrovarsi presenti, non si osserva perché per la poca autoritá ben spesso toccherebbe la spesa al collegio. Di che nasce che cosí li Chirurgi nel medicare, come li Barbieri nel cavar sangue non avendo mai visto situazione di membra, arterie, vene e nervi e lochi loro fanno errori atroci, tagliando una cosa per un'altra o simili con danno estremo o irreparabile delli pazienti" (ASR, *Università*, b. 61, fol. 789v).

25. *Statuta 1531*, chapter 64.

26. This is a manuscript copy of the reform of the statutes promoted this time by Antonio Porta, *protomedicus* in 1584. It is housed in ASR, *Università*, b. 61, pt. IV, fols. 1–71 (henceforth *Statuta 1584*). This arrangement in eight points disappears beginning with the 1676 statutes.

27. *Statuta 1531*, chapter 64; *Statuta 1582*, chapter 53; and *Statuta 1584*, chapter 53.

of the political (the cardinal vicar) and judicial establishments (the senator or governor of Rome). Although it was the *protomedicus,* joined by
his counselors, who sanctioned it, only the vicar had the authority to put
it into effect, on the basis of criteria that are unclear from the surviving
documentation. One could make some hypotheses concerning the cardinal vicar's role. He functioned as bishop of Rome and thus had broad
jurisdiction over the territory in question, and he was probably able to
specify the time, place, and appropriateness of the anatomy, as well as the
criteria for the selection of the cadaver. It may also be that although it is
specifically mentioned in the statutes, the cardinal vicar's role was more
formal and symbolic than materially effective. His mention in the regulatory text that sets out the various phases of the anatomical ceremony
indicates the involvement of papal participation in the religious and political aspects of this operation.

The responsibility for procuring the cadavers for the dissections fell on
the senator or the governor of Rome. The two tribunals that enjoyed the
broadest jurisdiction in the city and were responsible for cases involving
public order and criminal law reported to them. The governor, unquestionably, was the principal figure in the adjudication of criminal cases in
Rome.[28] The fact that those two officials supplied the cadaver seems to
imply that the body was that of someone who had been sentenced to death
by one of the two tribunals that had the authority to dispose of it. The statutes of Padua are much more explicit with regard to cadavers that could
be dissected, declaring suitable for dissection "any cadaver of whatsoever
criminal, condemned by the *praetors* to capital punishment."[29] The Pisan
statutes are equally straightforward on this matter: "We order that the
Lord Commissioner of Pisa assign to the Rectors the cadavers of delinquents sentenced to death, at every request on their part."[30]

The determination of which cadavers would be supplied to the university
thus falls, yet again, to the discretion of the papal state apparatus. I shall
consider the specific criteria employed in the selection. This referral to the

28. On this position, see N. Del Re, *Monsignor Governatore di Roma* (Rome, 1972).

29. *Stat. Gymn. Pat.,* l. II, chapter 28. On the Faculty of Arts in Padua, see also G. F.
Tomasini, *Gymnasium Patavinum* (Udine, 1654) and J. Facciolati, *Fasti Gymnasii Patavini*
(Padua, 1757).

30. This is the decree by which Cosimo I de' Medici established the chair of anatomy in
Pisa, which was later inserted in the statutes of the *Studium* (chapter 1). cf. A. Fabroni,
Historiae Academiae Pisanae, 3 vols. (Pisa, 1791–95), 2:73, note 1. In Bologna, instead, the
1405 statutes provided that the doctors themselves should acquire the cadavers for dissec-
tion. After a reform of the statutes in 1442 it became the responsibility of the *podestà* of
Bologna or of his substitute (see C. Malagola, ed., *Statuti dell'Università e dei Collegi dello
Studio bolognese,* [Bologna, 1888], p. 289). See also L. Simeoni and A. Sorbelli, *Storia dell'-
Università di Bologna* (Bologna, 1940) and G. Ferrari, "Public Anatomy Lessons and the Carnival: The Anatomy Theatre of Bologna," *Past and Present* 117 (1987): 50–106 at 53–55.

highest criminal authorities and the attention of the cardinal vicar are extremely significant and are an indirect manifestation of the rigorous precautions established to safeguard the procedures preliminary to dissection.

The third item in the Roman statutes specified that the college, to avoid any possible confusion, should establish by decree the procedural methods and the opportune times for the anatomies.[31] The available documentation is silent on this point as well. No trace of these decrees has survived in the deposit *Università* of the *Archivio di Stato* in Rome, so a reconstruction can only be attempted on the basis of indirect evidence. What are the features in a public anatomy that could cause "confusion?" Is this an allusion to irregularities in procedure or to controversies that could arise from it? One of the elements to which these decrees must refer is explicitly mentioned: "according to the appropriateness of the time." Others could pertain to the selection of the site at which the dissection was to take place, the duration of the ceremony, the qualifications and number of persons admitted, the "tax" paid, and, quite probably, other details concerning the condition of the cadaver and how it should be used.

We should examine some of these items briefly; others can be considered more fully later. The Roman statutes gloss over certain particulars that are treated in detail in the regulations of other universities and that imply the necessity of establishing more detailed "categories" (*facere capitula*). An instance of this glossing over can be seen in the statutes' instructions pertaining to the choice of the best time to hold the public dissection: in Rome, this choice was left to the discretion of the *protomedicus,* with no further qualifications, except that it had to be suitable for the season. The Pisan statutes are much more specific, giving "wintertime"[32] as the period during which the dissection should be conducted, and the Paduan ones stipulate that it should be held "after studies have begun and before the end of February."[33] In Bologna, the first statutes (1405) do not impose any special time limit, although the great majority of public dissections were held in January and February, especially during the Carnival holidays.

This latter custom became law in 1570 with a decree, introduced in the 1602 statutes, establishing that dissections could be performed from the end of the first trimester, precisely during the Carnival vacations.[34] In fact,

31. This point is stressed in all the statutory reforms of the *Studium Urbis* until 1676. See *Statuta 1676,* chapter 53.

32. A. Fabroni, *Historiae,* 2:73, note 1.

33. *Stat. Gymn. Pat.,* l. II, chapter 28.

34. ASB, *Assunteria di Studio,* b. 2 and ASB, *Senato, Partitorum,* XXIII, fol. 32. This information on the University of Bologna is taken from G. Ferrari, "Public Anatomy Les-

even in Rome, throughout the sixteenth century, dissections were held primarily between early January and late February, with rare exceptions.[35] The custom of fixing the public ceremony of the unveiling of the cadaver during the Carnival vacations was thus tacitly respected even at the University of Rome. The choice of the period of Carnival as the proper time for the dissection had a twofold justification: first of all, it was the coldest time of the year, and thus the period when it would be possible to best preserve the cadaver; second, since it was a time when the university was on vacation, *lectors* and students, free of their courses, could attend an event that was exceptional from a didactic and sensational point of view. The season was suitable for yet another reason. The scheduling of the anatomy lesson during this period would give it a cultural as well as a practical connection to the Carnival. Carnival, as is well known, is the only clearly defined time of the year during which certain behavior (generally considered transgressive) is permitted by virtue of an implicit social pact. If the dissection was considered macabre and sacrilegious in certain respects and was looked upon as barely permissible—an act that avoided prohibition only by being circumscribed by certain regulations and by a rigorous ceremonial—then to perform it during Carnival implied that it too, along with other transgressive practices, was temporarily channeled into the sphere of the licit by its ritualization. Even if dissection was to be considered a transgressive and profaning act, it was tolerated because it took place at a time when every form of subversion and inversion was concealed under the guise of performance.[36]

---

sons," pp. 54, 64–66, 68, 71. The author adds that the dissections were here classified as "extraordinary" lectures and were thus held during the period when the regular (*ordinarie*) lectures were adjourned (ibid, p. 54, note 20).

35. The evidence on Roman dissections is taken from the *Libri del Provveditore* of the Archconfraternity of San Giovanni Decollato housed in the ASR. On the role of this charitable group in the history of Roman dissection and on the importance of this source for the present research, see the subsection of this chapter entitled "The Selection of the Cadaver: Explicit Criteria and Implicit Caution," below.

36. If it is true that the word "Carnival" originates in the Latin "caro," as has been suggested by Peter Burke in his *Popular Culture in Early Modern Europe* (London, 1978), and if the several Carnival practices are indeed linked to flesh in its multiple aspects (food, sex, violence), it may not be too daring to suggest that dissection, practiced during the Carnival period, fits rather appropriately into those activities connected with sensuality characteristic of that time of year. For the etymology of "Carnival," see also J. C. Baroja, *El Carnaval: Analisis historico-cultural* (Madrid, 1965) (French translation: *Le Carnaval* [Paris, 1979], pp. 30–49). On the relationship between Carnival and dissection, Ferrari's article mentions the possibility of a "Bachtinian analysis of the way the university institutions responded to public anatomy lessons" (p. 104). The reference is naturally to Bakhtin's *Rabelais and His World,* trans. H. Iswolsky (Cambridge, MA, 1968) and to his *Problems of Dostoevsky's Poetics* (Minneapolis, 1984). Ferrari's argument moves along parallel lines

We know little, on the other hand, about the location in Rome where the anatomy lesson took place. It is possible to hazard a guess on the basis of indirect and not always verifiable evidence. Bartolomeo Piazza, at the end of the seventeenth century, talks about dissections taking place in the hospital of the *Consolazione,* in the hospital of *Santo Spirito in Sassia,* in that of *San Giacomo degli Incurabili,* and, finally, beginning in the second half of the century, in the anatomical theater constructed in the Sapienza.[37] There is no doubt, however, that from the sixteenth century onward, even though a theater had not yet been built, some dissections took place in the rooms of the *Studium.*[38] They were presumably held in the chapel that frequently served as a lecture hall, even though it continued to have a preeminently religious function,[39] or in some other hall con-

focusing on two points: on the one hand, on the ritual and festive character of both events and the alleged response of the public to the two phenomena, and, on the other, on certain aspects common to the "grotesque body" characteristic of the Carnival and to the dissected body, as they can be deduced from a reading of Rabelais. This is intended to prove the existence of an attitude shared between the "popular" sense toward the "grotesque body" within the Carnival ritual and the institutional approach to the dissected body in the anatomical ceremony (and this included physicians, students, and simple bystanders present at the demonstrations). Ferrari's otherwise careful and convincing study lacks any analysis of the dialectic transgression/tolerance that, especially in the first half of the sixteenth century, appears to me to underlie both phenomena and their temporal association.

37. "Spettacolo in vero curioso e magisterio sensibile dell'umane miserie è quello, che si costuma fare in una gran sala del Ven. Ospedale della Consolazione nel giorno del suo titolare della Natività di Maria Vergine, ove si mette in pubblico cospetto, et in scheletri spolpati l'ossatura, il sistema e l'ammirabile Architettura del Corpo Umano, con la relazione de' nomi e copioso vocabolario delle parti di esso, con tutti li stromenti dell'Anatomia . . . Una simile Scuola d'Anatomia quasi ogn' anno nell'Inverno si fa sopra un corpo umano nel grande Archispedale di San Spirito in Sassia, insegnandosi dai più eccellenti Chirurghi et imparandosi dai più Studiosi Scolari il modo di conservare la vita ai vivi dall'estinta dei morti . . . s'aprí parimente questa nobile Accademia della Notomia, col suo artifizioso Teatro per il commodo universale de risguardanti per ordine della San[ta] Mem[oria] d'Innocenzo Undicesimo non meno paternamente intento alla salute dell'anime che a quelle dei corpi" (see B. Piazza, *Eusevologio romano overo delle opere pie di Roma . . . con due trattati delle Accademie e Librerie celebri di Roma* [Rome, 1698], pp. 23–26, chapter 8: "Dell' Anatomia"). This anatomical academy was founded in 1675 by the Roman physician Guglielmo Riva, according to the 1676 statutes of the Roman College of Physicians. And in the dedicatory letter of the statutes we read: "Crebrescunt passim tam publicae in Xenodochiis, quam privatae domi D. Guglielmi Rivae Chirurgi in Urbe Clarissimi Anatomicae corporum dissectiones praeclaro ad indagandas morborum sedes rudimento" (fol. 3r).

38. See ASR, *San Giovanni Decollato,* b. 2, l. 4, fols. 29v and 163v; b. 4, l. 9, fols. 64v and 100v.

39. In this regard, see P. Capparoni, "I maestri d'anatomia nell'Ateneo romano della Sapienza durante il sec. XVI," *Bollettino dell'Istituto Storico Italiano dell' Arte Sanitaria* 6 (September–October 1926): 201. The author suggests that the chapel of the *studium* was used as an amphitheater. He also maintains that some dissections were being held publicly at Castel Sant'Angelo: "I recall that at the time of the restorations being carried out by General Borgatti at the Castel Sant'Angelo in Rome, the word *"ANATOMIA"* written in

veniently equipped for the occasion "with instruments, stools, tables, incense, sponges, rose water, vinegar." It seems probable that public anatomies in hospitals came later, along with a larger selection of cadavers and a partial liberation from the existing regulations.[40] At this stage there were simply locations temporarily employed as theaters for dissection, evidence of the hesitation to provide a stable environment for a method of teaching and a discipline that had not yet achieved an autonomous and formalized status.

The University of Padua was the first to have a permanent anatomical theater built in 1594, thanks to Fabrizio d'Acquapendente. The university certainly offered a more detailed and organized system than the one outlined in the Roman statutes. "So that the matter may be properly ordered and proceed with maximum utility," the rector was to elect two "suitable" students who had been enrolled in medicine for at least two years and who had participated in other public anatomies. These were called *massarii anathomiae.* They were obliged to make the appropriate arrangements for the location of the anatomy lesson and to prepare the instruments and all that was required in the operation. They had to establish the fee to be paid by all those who wished to attend the lesson ("the tax to be in proportion to the accompanying expenses"). In Rome this function was in the hands of *depositarii,* whose responsibility, however, was limited to the bookkeeping aspects: they had to fix a "tax," keep count of the income and expenses of the demonstration, and, as has already been mentioned, make arrangements so that whatever money remained was distributed to the poor for the salvation of the dissected cadaver. The *massarii* also controlled the admission of the public to the anatomy lecture. The Paduan statutes strenuously emphasize this point: "They are charged with overseeing that no student, unless he has matriculated and has been enrolled in medicine for at least one year, should be admitted. The Lord Rector, with one associate, all doctors who serve as *lectors,* and all doctors of the college, the two *massarii* themselves, as well as two other poor students

---

black with lettering of the second half of the sixteenth century was found in the courtyard of Alexander VI underneath the whitewash and stucco. The word was repeated on the wall over the doors of a few rooms which opened on the courtyard" (p. 203).

40. In Padua, for example, where the first "university" dissections of which there is a trace date to February 8 and April 4, 1430: see Leonardo da Bertipaglia, *Recollectae habitae supra quarto Avicennae,* in Guy de Chauliac, *Ars chirurgica* (Venice, 1546), fols. 261v–302r and 299v. Cited in G. Ongaro, "La medicina nello studio di Padova e nel Veneto," in *Storia della cultura veneta,* ed. G. Arnaldi and M. Pastore Stocchi (Vicenza, 1981) pt. 3, 3:95–96. We have information about dissections in hospitals only from 1578 (J. J. Bylebyl, "The School of Padua, Humanistic Medicine in the Sixteenth Century," in *Health, Medicine and Mortality,* ed. C. Webster [Cambridge and New York, 1979], p. 350).

. . . and elected counselors are to be admitted without payment. All the rest are to be rejected. Neither the Rector, nor the counselors, nor the *massarii* have the authority to admit anyone who is not matriculated, who has not studied medicine for a year and who has not paid."[41] The organization of the public anatomy lesson in Pisa resembled the Paduan one: in order for anyone to be admitted the same requirements called for by the latter statutes had to be satisfied. In Pisa two students, called *anatomistae*, were elected to carry out the same duties as the *massarii* fulfilled, and their principal purpose was to see "that everything is carried out in good order." In addition to guaranteeing the correct procedure of the anatomy lesson from the formal and organizational point of view, they had to ensure that "there should be no cause for tumult in the aforesaid anatomy."[42]

There is here a reiteration by the use of a synonym of that "confusion" mentioned in the statutes of the Roman physicians. "Tumult" and "confusion": these two words, inevitably, invoke the scuffles that could break out over an unsuitable choice of cadaver. These disturbances might be provoked by relatives and friends who protested the profanation of the remains of their loved one, by spectators at dissections, by barbers and surgeons over the exorbitant entrance fees, or by the controversies that could arise from the assignment of places normally distributed on the basis of the position, seniority, and "dignity" of those present.

The conventions of the Roman statutes ("so that confusion may be avoided") had the same purpose as that attributed to the actions of the Paduan *massarii* and the Pisan *anatomistae*: to organize every part of the demonstration in its strictly technical and ceremonial aspects down to the smallest detail so that it might be conducted in absolute conformity to the statutory norms, and equally to what might be called the cultural ones, in order to avoid misunderstandings over the profane employment

41. *Stat. Gymn. Pat.*, l. II, chapter 28.
42. A. Fabroni, *Historiae*, 2:73–74, note 1. Even in Bologna there were *custodes* assigned to hold back the importuning populace (*importunam plebem*) (A. Benedetti, *Anatomice sive historia corporis humani. Ejusdem collectiones medicinales seu aphorismi* [Venice: Bernardino Guerraldo Vercellensis, 1502], fol. 9r). On the subject, there is a decree of 1586 by Cardinal Salviani in Bologna, translated by Ferrari: "The stated aim of the decree was to 'prevent the disorders occasioned by the anatomy, and to ensure in future that it may be followed with the proper calm and usefulness to those attending in order to learn' and 'so that it may be heard and seen with due decency (*modestia*).' It provided for the appointment by the university of four particularly sober scholars, whose task would be to stand at the door of the theatre when there is an anatomy lesson, and to allow and refuse entrance as they see fit, in order that the theatre accommodate only doctors, scholars and other persons of good quality, who enter therein so that they may hear and learn, and not create an uproar, as sometimes occurs, with the prohibition of payment for entrance made by scholars to whomsoever" ("Public Anatomy Lessons and the Carnival," p. 70).

of the body, which might cause unseemly outbursts. Equally, the correct regulation of this academic ritual promoted the didactic efficacy of the public anatomy demonstration.

THE ANATOMY LESSON: SOME HISTORY

In Rome it was the College of Physicians that arranged for the election of the person or persons who would provide the reading at the public anatomy lesson. The 1531 statutes, in fact, specified that the doctors of the college themselves should decide on the appropriateness of assigning responsibility for each individual cadaver to one or more *lectors*.[43] It was also the college that selected the two *censores*, one of whom would serve as *ostensor* (demonstrator), the other as *incisor* (dissector). The Pisan statutes are rather offhand regarding the distribution of these tasks, but the recommended text and the pace of the reading are minutely described in the Paduan ones: "for rector and counselors someone should be selected from among the temporary lecturers to recite and read from the text of Mondino's anatomy, and another from among the regular physicians, whether practical or theoretical, should solemnly enunciate the aforesaid text, and what he declares according to that text should be verified to the naked eye on the cadaver. Nor should another item be read or demonstrated, without the preceding having been read and demonstrated, providing the reading accompanying the incisions and dissections."[44]

At Padua, as in Rome, three people worked together in the anatomy lesson and in the dissection. What varied, insofar as this can be inferred from the statutes, was a hierarchical shift between *lector, ostensor,* and *sector*. From the Roman regulations, it seems that the person of greatest prominence in the demonstration must have been the *lector,* who as reader coordinated the entire didactic/theatrical operation. The Paduan statutes, instead, underline the fact that the *lector* was to be chosen from the "Extraordinary" faculty, in other words from among those whose university status was unquestionably lesser in terms of prestige and remuneration.[45]

43. The 1676 statutes provided for the election of two docents, who, on alternate days, would read and explicate the anatomy texts; alternatively, the entire assignment could be entrusted to one person: "Item Lectores eligantur, qui alternis diebus interpretentur: vel unus, qui interpretandi onus totum suscipiat: Adhaec duo eligantur, Alter ad incidendum, ad ostendendum idoneus Alter" (*Statuta 1676,* chapter 51).

44. *Stat. Gymn. Pat.,* l. II, chapter 28. The text continues: "[Q]uod nullus doctor quicquam dicere audeat nisi postquam scholares particulam viderint. Dum vero altera inciditur, super praecidenti iam visa quilibet doctor dicere, et proponere possit ad scholarium utilitatem, quod sibi videbitur."

45. The Paduan university system, which closely resembled that of much of the rest of Europe, provided for each of the principal instructional subjects (practical medicine, theo-

The *lector*'s job was to read the *Anatomia Mondini*. The most prominent figure in this case was a "public" reader (it did not matter whether he was in practical or theoretical medicine).[46] His responsibility was to clarify and to comment on the passages read and then to demonstrate and affirm his commentary on the cadaver; the last person in this implicit hierarchy was the one physically entrusted with the dissection. The discrepancy between the arrangements of the two universities on the assignment of roles in the public anatomy lesson, although noteworthy, does not substantially affect its most significant feature: in both cases the anatomy lesson was carried out in accordance with what, in the previous chapter, is identified as the quodlibetarian model, in which the roles pertaining to practice are kept distinct from those pertaining to theory. In both cases the dissection is entrusted either to a barber or to a surgeon, not only because of their specific competence but also by virtue of the fact that manual work was held in low esteem. In spite of the innovative suggestions made by Vesalius and of the publication of numerous anatomy texts intended for university teaching, this model survived for most of the sixteenth century, and Mondino's *Anatomy* continued to function as the textbook that accompanied and guided dissection in Padua.

The chair of anatomy is listed only in 1552 in the rolls of "public" *lectors* at the University of Rome, and, presumably, at that time it was vacant, since no name is entered for that discipline.[47] It then reappears as an autonomous chair in 1583, when it is assigned to Arcangelo Piccolomini, who held the chair of practical medicine contemporaneously and received a differentiated stipend for the several positions entrusted to him. It is clear, however, that public anatomies were performed in Rome even before a permanent chair was established, as occurred in Padua where, although a course of lectures on anatomy (in association with surgery) was not established before 1584 for Girolamo Fabrici of Acquapendente, there are records of such lectures for some decades earlier. Sporadic mention of

---

retical medicine, etc.) five positions decreasing in order of prestige and corresponding salary: first and second Ordinary lector, first, second, and third Extraordinary lector (see J. J. Bylebyl, "The School of Padua," pp. 343–44, and Tomasini, *Gymnasium*, pp. 291–332). On this point, see *Statuta Patavini*, fols. 34r, 37v–38r, 46v.

46. "These designations are somewhat misleading, however, since both dealt with a combination of theoretical and practical issues, and neither was practical in the sense of 'clinical'" (see J. J. Bylebyl, "The School of Padua," pp. 338–39). On the inappropriate distinction between theory and practice, as far as Padua is concerned, see S. Santorio, *Commentaria in primam fen primi libri Canonis Avicennae* (Venice: I. Sarcinam, 1626), coll. 4–6, 37–39.

47. ASR, *Università, Cimeli*, roster of lectors for the year 1552. Lists of this sort are rare for the entire sixteenth century. See E. Conte, ed., *I maestri della Sapienza di Roma dal 1514 al 1787: i Rotuli e altre fonti*, (Rome, 1991).

dissections goes back as early as the fourteenth century, although they cannot be defined as "public anatomies" in the proper sense of the term.[48]

Dissections were certainly taking place in Rome at least from 1531, the date of promulgation of the statutes of the College of Physicians, which established the regulations to be followed for the public demonstrations of anatomy. But, as early as March 22, 1512, the cadaver of one Giovanni da Monte was turned over to the physicians and students of the *Sapienza* to be used in anatomy lessons: this is the first documented instance of a dissection occurring in the papal city.[49] It is not until February 14, 1518, that the next piece of evidence of another dissection performed by the doctors of the *Studium* turns up, and then it is not until 1522 that any of the sources again mentions a cadaver that is handed over to physicians "to perform an anatomy."[50] It is clear that the absence of a chair specifically allocated to anatomical study did not imply that the discipline was not taught before it was formally incorporated.

With the chronological boundaries established, the question still remains: who precisely in the university performed the dissections, as *lector* and *sector*? At this point it may be opportune to set out a short chronology of those who taught anatomy in Rome, with or without a specific chair, from the end of the fifteenth century to the beginning of the sixteenth. Gabriele Zerbi of Verona, who taught medicine and philosophy in Bologna between 1475 and 1483 before moving to Rome, where he spent approximately a decade, was one of the early prominent figures in the anatomical field. He was the author of a *Liber anathomiae corporis humani,* published in 1502 in Venice, in which he recommends dissection as an essential tool for diagnostic medicine and surgery.[51] Matteo Corti, papal physician and a celebrated anatomist active in Rome at the time of Clement VII, was another champion of demonstrations performed directly on the cadaver.

In 1539 Paul III summoned the Neapolitan Alfonso Ferri to Rome to

48. "La medicina nello studio di Padova," pp. 89–99. The author cites a decree of the Venetian Great Council, dated May 27, 1368, enjoining the College of Surgeons to carry out dissections on the cadaver (AS Ven., *Maggior Consiglio. Deliberazioni,* 19 *Novella* (1350–84), fols. 114v–115r).

49. ASR, *San Giovanni Decollato,* b. 1, l. 1, fol. 31v.

50. The dissection was performed on the cadaver of a woman, Catherina di Lorenzo, as is recorded in ASR, *San Giovanni Decollato,* b. 1, l. 3, fol. 32v.

51. The dissections, he writes, should be performed on animals (preferably monkeys because of their resemblance to human beings), but also on men: the bodies of those who died suddenly were preferable, neither fat nor thin, neither old nor young. The treatise offers a summary idea of all the parts that make up the human body and, interestingly, suggests some rudimentary methods for preserving the cadaver from accelerated decomposition. See on the subject, A. Portal, *Histoire de l'anatomie,* 1:247.

serve as his physician and at the same time to fill the chair of surgery. His name appears in the rosters of the *Studium* for the years 1542, 1548, and 1549, although it seems that he taught there until 1561, the year he returned to Naples.[52] We do not know if he was a physician or surgeon by profession, nor on the basis of what qualifications he had been enlisted as a papal physician. According to Renazzi, however, although he does not cite a source, it is certain that it was Ferri who inaugurated a course of regular anatomical study at the University of Rome, a discipline that, at least initially, was intimately linked to that of surgery.

Realdo Colombo was active in Rome contemporaneously with Ferri. As already mentioned, he was at the University of Pisa in 1546 and was in Rome two years later as physician to Paul III[53] and to work on the *De re anatomica* in collaboration with Michelangelo.[54] It seems certain that Colombo taught at the University of Rome for over a decade, as he himself affirms in the text, and his sons in the preface of the treatise,[55] and as is also demonstrated by the numerous references to public anatomies performed "in the Roman Academy" reported in the same work.[56] Juan Valverde, in his letter of dedication to Philip of Spain contained in the Italian translation of his anatomical work and dated May 20, 1559, speaks of Realdo, moreover, as an "excellent anatomist and my teacher in this faculty."[57] Realdo Colombo's name does not appear in the faculty rolls, so that it is impossible to judge the type of lectorship to which he was assigned, or whether his appointment was as "ordinary" or "extraordinary."

In 1563 it was the turn of Bartolomeo Eustachio, who had come to Rome together with Giulio della Rovere at the end of the 1540s, to read practical medicine.[58] It is unclear when he received his first university

---

52. This last date can be calculated with the help of the rosters: in 1552, as was mentioned earlier, the names of the *lectors* occupying the chairs of surgery and anatomy have been cancelled. The next surviving entries are for the year 1563, and here Alfonso Ferri is no longer the *lector* in surgery. The information is taken from Renazzi, *Storia*, 2:107–10. Ferri was also papal physician to Julius III and Paul IV.

53. G. L. Marini, *Degli Archiatri pontifici*, vol. 1.

54. See chapter 1, pt. 5.

55. R. Colombo, *De re anatomica libri XV* (Venice: Nicolò Bevilacqua, 1559), p. 139.

56. Moreover, on January 27, 1557, the Archconfraternity of San Giovanni Decollato entrusted the cadaver of Giovanni di Natale "to certain students at the Sapienza" so that Colombo could dissect it (ASR, *San Giovanni Decollato*, b. 2, l. 4, fol. 29v).

57. J. Valverde, *Anatomia*, fol. 2v.

58. ASR, *Università, Cimeli*, roster of lectors for 1563. Eustachio's *prosector* was Pietro Mattei, whose annual salary was fifty *scudi*. See P. Capparoni, *I maestri*, p. 203, and also *Memorie e documenti riguardanti Bartolomeo Eustachio pubblicati nel quarto centenario della nascita* (Fabriano, 1913).

chair, but it was certainly not before 1552 or after 1556, the year in which Valverde alludes to Eustachio as his teacher at the *Studium* of Rome. In the meantime, he performed instead dissections, or rather, autopsies, for the hospitals of *Santo Spirito in Sassia* and of *Santa Maria della Consolazione*, both centers for trauma. His responsibility was often to check the cause of a patient's death and of the death of all those others who passed through the hospitals, providing an expert opinion to the judicial investigations even after death. It would seem that he had established a sort of anatomy theater that permitted a larger public to attend the autopsies that were becoming true lessons in anatomy. Eustachio continued his dissecting activity even during the period when he taught at the *Sapienza,* and from this came his *Opuscula anatomica*[59] and the forty-seven copper engravings drawn by Eustachio himself and by Pier Matteo Pini and etched by the Roman Giulio de Musi.[60] By 1567, however, Eustachio's name no longer appeared on the faculty rolls, although he continued to live in Rome, keeping his place in the College of Physicians until his death in 1574.[61]

Benalba Brancalupo, a teacher of surgery from 1563, was entrusted with the public anatomies from at least 1568.[62] He kept his chair of surgery until 1583 but despite his efforts did not retain his lectorship in anatomy for any length of time, because his students were unhappy with his teaching. Their complaints are documented in the records kept by the head beadle for the benefit of the papal authorities on the progress of instruction and on the mood within the *Studium*. The complaints had initially been based on the fact that Brancalupo read in the vernacular rather than in Latin,[63] but during the 1568–69 academic year the criticism became more pointed and harsher. That year no anatomical demonstra-

---

59. Published in Venice in the printing shop of Vincenzo Luchino in 1563 and reissued the following year.

60. They were published two centuries later by Lancisi with the title *Tabulae anatomicae Bartholomaei Eustachii, quas e tenebris tandem vindicatas . . . praefatione, notisque illustravit ac publici juris fecit Jo. Maria Lancisius* (Rome, 1714). These charts had been completed in 1552, as is stated in the *Opuscula,* chapters 13 and 16. See also, on this work, G. Petrioli, *Corso anatomico o sia Universal commento nelle tavole del celebre Bartolomeo Eustachio* (Rome, 1742), and, by the same author, *Le otto tavole anatomiche con cinquanta figure in foglio delineate per compimento dell'opera sublime et imperfetta del celebre Bartolomeo Eustachio* (Rome, 1750).

61. ASR, *Università,* tome 48, fol. 72r–v. On September 13, 1574, the day Eustachio died, Girolamo Cardano was elected to his place in the college.

62. Normally Brancalupo's salary for teaching surgery was 100 *scudi*. In the 1568 rosters the figure had doubled because he had begun his regular instruction in anatomy.

63. During the 1567–68 academic year the beadle noted alongside Brancalupo's name: "non satis colit linguam latinam" (ASV, Arm. XI, t. 93a, *Relazione del Bidello,* 1567–68).

tions were held at the university and the students "loudly lamented"; they attributed the reason to the "inexperience" of Benalba Brancalupo and asked that someone else be hired to "read surgery" who would "perform anatomical dissections," emphasizing the scientific interest and the absolute necessity of the dissection of cadavers for the proper care of the living.[64] The following year Brancalupo's teaching again provoked the protests of his students, who by now were greatly reduced in number, and the dissections required by the course of study were not held, causing, as the beadle commented, great inconvenience to the students and even greater harm to the sick.[65] In the report for the academic year 1570–71, next to the name of Brancalupo, *lector* in surgery, the beadle noted that "He wanted to do the anatomy assigned to him. He did not know the Latin language sufficiently and read his lecture in the vernacular. Master Scipio who is serving at the hospital of *Santo Spirito* would have been better." That year the public anatomy lecture was being conducted, to the students' great acclaim, by Francesco Ginnasio (Gymnasius or Genesius in the manuscripts) who held the chair of theoretical medicine.[66] In 1572 Gregory XIII summoned Costanzo Varolio from Bologna to try to correct the sad state into which the teaching of anatomy had fallen.[67] The dissections for the years 1573, 1574, and 1579 were entrusted to Scipione de' Rossi (the "master Scipio" mentioned by the beadle) who was given the title of *incisor*.[68] In January 1575 it was still Varolio who taught the students anatomy,[69] while just two weeks later, over the protests of the students, Brancalupo performed an anatomical demonstration, but one limited to the head.[70] Anatomical exercises were suspended during the following two years. Between the 1570s and the 1580s Pietro Crispo, who held the position of Ordinary of theoretical medicine in the *Studium*

64. Ibid, 1568–69 year.

65. Ibid, 1569–70 year.

66. "Magister Franciscus Genesius; hic benemeritus laudatur a scholaribus quia aptissimus esset ad Anathomiam facendam" (ibid, 1570–71 year).

67. His name never appears in the roster of university faculty, but the epitaph on his tomb is unambiguous: "Medicinam et Chirurgiam percallens—Extrahendi calculi peritissimus—Cum in Gymnasio Romano—Anatomicam lectiones—Sectionemque profiteretur" (from P. Capparoni, "I maestri d'anatomia," p. 206).

68. ASR, *San Giovanni Decollato*, b. 4, l. 8, fol. 95v; b. 4, l. 9, fols. 100v and 102v. In addition to his clinical activity, Scipione served as papal surgeon under Julius III, Paul IV, and Pius IV. In a note from the beadle to papal authorities written in 1566, we read: "Il Chirurgo si grava di aver carico della Anatomia senza argomento, questo huomo molto assiduo al leggere, et son più anni che medica in S.to Spirito." This attests to his participation in the life of the *studium* for fifteen years or so without his ever having occupied a chair.

69. ASR, *San Giovanni Decollato*, b. 5, l. 10ff., fols. 3v–4r.

70. The skull was that of Paolo Buscatti, who had been condemned to death by the governor. See ASR, *San Giovanni Decollato*, b. 5, l. 10, fol. 7r.

Urbis (teaching in competition with Francesco Ginnasio), lectured on anatomy.[71]

At the beginning of the 1580s the anatomy lesson was being coordinated by Arcangelo Piccolomini,[72] who was assigned the reading of the text and was assisted by Scipione de'Rossi, who "most diligently dissected and demonstrated." From this it would appear that, starting from this date, the separation of roles enjoined by the statutes was no longer being respected, with the function of *ostensor* and that of *sector* (or *incisor*) essentially being carried out by the same person. In 1582 Arcangelo Piccolomini became Ordinary *lector* in practical medicine at a salary of 300 *scudi* per annum. He then became a *lector* in anatomy, which at that time was considered a separate chair, with a stipend of 100 *scudi*. His dissector was Leonardo Blandino, with a salary of 40 *scudi*. We do not have any information on who Piccolomini's demonstrator might have been; perhaps it was Blandino or the *lector* in surgery (Benalba Brancalupo). Finally, between 1587 and 1619, the joint chairs of surgery and anatomy were held by Angelo Antonini.

The *lectors'* rolls bring clearly to light two elements concerning the anatomists, or, more accurately, the persons who performed the public dissections. The first concerns their recruitment and their academic and professional identity: they were frequently non-Roman physicians who had already achieved a certain academic and professional notoriety elsewhere and who had been called to Rome as doctors to the pope and his family. Some of them became members of the college and senior general physicians (*protomedici*), so that all the highest positions to which a physician could aspire in the papal state were united. In the *Studium* they always lectured as well on another subject with which the teaching of anatomy was associated (practical or theoretical medicine, surgery). The *sectores,* for example Scipione de' Rossi and Leonardo Blandino, performed also as physicians, surgeons, and barbers at a Roman hospital, especially at *Santo Spirito.*[73] Their respective careers thus appear quite

71. His anatomical activity emerged from a letter that the medical students of the University of Rome addressed to the supervising commission of cardinals toward the end of the sixteenth century, petitioning for "un lettore [of anatomy] conforme al bisogno . . . a utilità del publico et delli Scolari" (ASR, *Università*, t. 69, fols, 1v–2r). Crispo was also general *protomedicus* in 1577 and 1582, as well as personal physician to Pius V.

72. Ferrarese in origin, he had taught philosophy at Bordeaux and was called to Rome by Pius IV to become his personal physician. He was a member of the college and became *protomedicus* in 1582; he was also the author of an anatomy treatise (*Anatomicae Praelectiones*[Rome: Bonfadini, 1586]) containing a few illustrations.

73. In fact, more than *lectores,* as has been maintained, this hospital furnished *sectores* to the university: "gli incisori da tempi immemorabili li ha somministrati S. Spirito" (ASR, *Università*, t. 69, fols. 17 ff.: *Sistema per la catedra Anatomica,* 1746).

distinct: the latter had a clinical and practical education and had worked in hospitals; the *lectores,* instead, were physicians of great reputation, apparently more suited to the speculative life, to diagnostics and pharmacy, who had gone through all the stages in the *cursus honorum* to which a physician could aspire before coming to teaching.

The other element that seems important is the delay with which anatomy became an autonomous area of instruction, unyoked from the other disciplines, and self-contained. Although its evolution can only be traced hazily, in the documents available it always remains an auxiliary field: the chair of anatomy appears in the surviving registers of readers on three occasions only: in 1552 with the name of the incumbent left blank; in 1575 as part of the surgeon's instruction; and in 1583 given to Arcangelo Piccolomini, a former teacher of *medicina practica.* In Rome, unquestionably, throughout the sixteenth century, anatomy remained of secondary concern among academic offerings. Its lesser prestige in the academy is also revealed by the fact that the public demonstrations frequently failed to be held, not because of any lack of cadavers to dissect, but because of the absence or inadequacy of the academic personnel. In any case, as I have pointed out before, teachers of anatomy always taught some other subject of more established academic status. Even the salaries for the teaching of anatomy were notably lower compared to those of the other disciplines, compelling instructors to hold more than one chair. But this impression is qualified by other evidence attesting to the importance of anatomy and to the autonomy, quality, and continuity of its teaching. One need only recall the loud protests of the students in the years when the discipline was being taught by Benalba Brancalupo and the pressure exerted on the matter by the beadle.

The Selection of the Cadaver: Explicit Criteria and Implicit Caution

The statutes of the Roman College of Physicians provide no particular clues about the criteria that were to be followed in the selection of cadavers to be dissected. The only condition, as we saw earlier, was that the body had to be that of someone who had been condemned to death, since it was turned over to physicians by the governor or the senator.

In fact, as we shall see below, the entire history of western anatomy is marked by this tenacious association between dissection and the condemned, which persisted from Herophilus to Vesalius and beyond. Long before anatomical practice was governed by regulations written into university statutes, the use of the cadavers of the condemned in public dissections had tacitly been accepted. When these conventions were finally formalized, the custom was continued and became even more firmly established: the statutes appealed directly to the civic and judicial authorities for their supply of cadavers, requested for the university through the rector, counselors, *protomedicus,* or the entire college. The *podestà* in Bologna, the *commissarius* in Pisa,[74] the *praetors* in Padua,[75] the cardinal vicar and senator or the governor at Rome acted as regulatory officers to prevent disturbances over the selection of unsuitable cadavers and to prevent physicians and students from recourse to such sacrilegious acts as exhuming the dead,[76] or appropriating and dismembering bodies that did not deserve such ignominious treatment. In addition, from the moment of condemnation the body of the executed person belonged to the ruler or, through him, to the judicial authorities.[77]

Even when the body of the condemned person fitted the requirements of the anatomist and the spectators, which was usually the case when the body was quite youthful, in good condition, healthy, and of a strong musculature, such as to permit a successful demonstration,[78] the moral quality of the body to be dissected had to be evaluated at the same time. The statutes repeat with unusual emphasis such terms as "criminals," and "delinquents," to designate those who, having been found guilty in a terrestrial court of law and executed, might pass under the instruments of the *prosector.* Their bodies, punished and damned, would continue in their agony even beyond life, since their souls would pay in the hereafter for the sins they had committed.

To require that judicial officials select the cadaver signified, for physicians, for the university, perhaps also for the entire community, entrusting

74. A. Fabroni, *Historiae,* 2:73, note 1.

75. " . . . teneantur Praetores dictorum locorum, non obstante decreto aliquo aut consuetudine, vel aliis quibuscumque ordinibus sub poena praedicta tale cadaver pro praedicta causa, ut supra rectoribus et scholaribus assignare" (*Stat. Gymn. Pat.,* l. II, chapter 28).

76. On this matter, in a large part of Italian territory, the legislation was extremely severe, as was the law: the death penalty was imposed for the desecration of graves, since "sepulcrum violatores, cum sacrilegi dicantur": (P. Farinacci, *Praxis et theoria criminalis,* 3rd ed. [Venice, 1603], *quaestio* 20, nn. 115 and 124).

77. Although he did not have at his disposal contemporary texts testifying to this usage, it is interesting to note that in the eighteenth century Cardinal Lambertini, a future pope, traced the legitimacy of the anatomy lesson to the fact that the prince could dispose of the bodies of the condemned as he saw fit (*Raccolta di alcune notificazioni, editti e istruzioni pubblicate dall'eminentissimo e reverendissimo signor cardinale Prospero Lambertini vescovo di Bologna* (Bologna, 1737), vol. 3: "*Notificazione XXIII: Sopra la Notomia da farsi nelle pubbliche Scuole,* p. 267). See, on the subject, G. Ferrari, "Public Anatomy Lessons," p. 60, which is also the source of this information.

78. This is how Benedetti puts it, *Anatomice,* fol. 101v. On Benedetti, see *Dictionary of Scientific Biography,* ed. C. C. Gillespie (New York, 1970).

the responsibility to persons who offered the greatest guarantee that only those whose crimes were absolutely flagrant—so great that an earthly verdict could not be overturned even in the heavenly court of appeal—[79] would be condemned beyond death. The status of these bodies nonetheless appears special: they were marginalized, banished from society, no longer the dwelling of the sacred; at the very moment of their sin it seems as if the people who later become the subjects of dissection lost, along with their lives, the dignity that the integrity of their remains conferred. The association of dissection with sin or transgression shows itself first in the surreptitious and irregular practices carried on at the boundary between the licit and illicit. Later, unable to evolve otherwise, such practices were codified in formally expressed and normative texts made necessary when they began to spread and gather momentum. The statutes show how anatomy advanced into the realm of the legitimate through formalization and ritualization.

Of all the forms of execution, simple hanging was the one preferred by anatomists for bodies that would be later dissected. This form of killing offered a better didactic experience, since it allowed anatomists to operate on a whole body, one not disfigured by the torture, punishments, or mutilation usually inflicted in every other type of execution. Given the chronic shortage of bodies for dissection, which was frequently lamented by physicians and students during the first half of the sixteenth century, it seems logical that they would seek undamaged bodies so that at the public anatomy lesson a complete academic demonstration could be performed on every single part and with the utmost clarity.[80]

There is one other culturally significant consideration concerning the hanged person destined for dissection. In Rome, but also in Florence and Siena beginning in the fifteenth century, hanging, at least in theory, was reserved for criminals from the lowest classes and for especially repugnant crimes (the alternative was beheading, which was reserved primarily for the nobility and for persons of rank). At the end of the sixteenth century the jurist Prospero Farinacci, in his *Praxis et theoria criminalis*—a manual of law and penal jurisprudence that discusses and compares the

---

79. I shall return to this point at the subsection entitled "Between Saying and Doing" in this chapter.

80. Only rarely do our sources mention the use in anatomical exercises of parts of the body obtained from bloody executions involving mutilation. In 1586 "si consegnò a m. Agustino Antonini come nothomista la testa e la spalla di Giuliano da Fara che fu squartato" (ASR, *San Giovanni Decollato*, b. 6, l. 12, fols. 232r–234v). See also the illuminated letter "O" in Vesalius's *De humani corporis fabrica* (either 1543 or 1555 editions), in which the executioner presents the head of a condemned man to a cherub, symbolizing an anatomy student (fig. 36).

teachings of the greatest Italian criminal lawyers—declared that hanging was the normal sentence for atrocious crimes against persons and property and that "certainly the punishment of the gallows was greater than the punishment of beheading, since it is more ignominious."[81]

On October 16, 1569, the Congregation of Roman Deputies "over the governance of the Gymnasium" issued a decree enjoining that public anatomies be conducted "on the bodies of Jews or other infidels who have been publicly executed."[82] This injunction supports the hypothesis that the anatomist's scalpel was to be used on the bodies of marginalized, ignoble, and despised people so as to avoid prejudicing, as much as possible, the sentiments of Christian piety and the practice of forgiveness.

If previous terrestrial condemnation and public infamy were essential elements in the decision to condemn a person to public dissection, other conditions, whether implicit or explicit, needed to be satisfied. In Padua, the statutes called for the dissection of two cadavers yearly, "one a man, the other a woman, at least one of them," to be selected from among the condemned, as long as they were not from Paduan territory or Venetians (which would have incurred a financial penalty).[83] Similarly, in Pisa, at least on paper, two bodies were to be consigned yearly to the *Studium* and, in the event that they were not always available in the city at the proper time, "the Rector should write to Florence to the appropriate criminal officials (*Dominos Octo Custodiae, et Baliae*) to that end, so that the aforesaid cadavers might be conveniently procured."[84] But there was a caveat: "Anatomy was not to be performed on the body of any Florentine or Pisan citizen, or of any Doctor or student."[85] In the case of Bologna, a 1442 decree reforming the statutes ordered that the bodies of the condemned intended for dissection should come from at least thirty

81. P. Farinacci, *Praxis, quaestio* 18, nn. 84 ff.

82. ASV, Arm XI, t. 93, fol. 48r. On this point Camporesi relates, without indicating a source, that Leonardo Fioravanti blithely acknowledged that he had practiced dissections on the living, but only on infidel Saracens (*Le officine dei sensi* [Milan, 1985], p. 163).

83. *Stat. Gymn. Pat.*, l. II, chapter 28.

84. A. Fabroni, *Historiae*, 2:73, note 1. A similar case occurred when Realdo Colombo was teaching in Pisa, before moving to Rome. On January 15, 1543, a woman, Santa di Mariotto Tarchi from Mugello, was sentenced to death by Cosimo I for murdering her two illegitimate infants by suffocation: "Dicta Santa di Monte Excellentissimis Ducis fuit missa Pisis, ut de ea per doctores fieret notomia." The woman's dissection, performed by Colombo, was a memorable event since her execution took place during her menstrual period, and it was thus possible to demonstrate the vessels through which the flow occurred. Valverde, who attended the procedure, recalled the episode, mentioning explicitly that Cosimo I had condemned her to be anatomized (*Anatomia del corpo humano*, fol. 90a). Cf. E. D. Coppola, "The Discovery of the Pulmonary Circulation: a New Approach," *Bulletin of the History of Medicine* 31 (1957): 53.

85. A. Fabroni, *Historiae*, 2:73–74, note 1.

miles outside the city.[86] A century later, in 1561, the regulation lapsed, and the use of the cadavers of condemned persons from the environs of Bologna was also permitted, probably because of the difficulties encountered in the procurement of bodies for the annual public dissections. The only concern expressed in this reform was that "the person must not be upright," or "of honorable birth."[87] Similar dispositions were also on the books in Florence, Genoa, and Perugia.[88]

In Rome no regulation alluded to restrictions of any sort in the selection of a cadaver to be dissected, except that it had to belong to a person who had been condemned to death. The rest was based on custom: the details contained in the statutes and decrees of so many Italian universities were superfluous here, but they were nonetheless operative.

The available documentation indicates that in sixteenth-century Rome all those destined for dissection because they had transgressed papal laws were foreigners, non-Romans, with the exception of Paolo Buscatti (or Biscatti) whose head was dissected between January 28 and February 2, 1575,[89] and of Rodolfo Barnabeo, hanged January 14, 1587.[90] The rest came from Spoleto, Florence, Forlì, Bergamo, Milan, and Treviso, or they were French, Corsicans and "Turks."[91] Of the thirty-one bodies dissected between 1506 and 1600 mentioned in the official sources, thirty were executed by hanging.[92] Of all those who left a will before dying, only one, Annibale Furlano of Parma, declared that he was owed a certain sum of

86. *Statuti dell'Università,* rubric 19, p. 318.

87. *Reformatio Statutorum Almi Gymnasii Bononiensis Philosophorum et Medicorum,* 1561, document C, appended to Corradi, "Dello studio e dell'insegnamento," pp. 33–35.

88. A. Gherardi, ed. *Statuti dell'Università e Studio di Firenze* (Florence, 1881), p. 74; P. L. Isnardi, *Storia dell'Università di Genova,* 2 pts. (Genoa, 1861–67). The 1481 statutes excluded the cadavers of persons "oriunda ex loco unde potestas traxit originem, nec Ianuensis aut de districtu" (1:229); G. Ermini, *Storia dell'Università di Perugia* (Bologna, 1947), p. 153.

89. ASR, *San Giovanni Decollato,* b. 5, l. 10, fols. 5v–7r, 8v.

90. Ibid, b. 7, l. 14, fols. 53v–54r.

91. See the table in the appendix with the listing of all executed persons who ended up on the dissecting table in Rome between 1506 and 1585. This information comes from the books of the *provveditore* of the Archconfraternity of San Giovanni Decollato preserved in the Archivio di Stato, Rome (on this institution and its role in the anatomical ritual, see the subsequent section). These books record the names of all condemned persons executed in Rome, with their testaments and a summary account of their fate from prison to burial. A simple listing of their names, arranged both alphabetically and chronologically, can be found in the volume entitled *Inventario dei condannati a morte* in the same archival repository. Unfortunately, no books are extant for the years 1522 to 1557, a crucial period for this study.

92. The single exception concerns Giuliano di Fara, who was quartered. ASR, *San Giovanni Decollato,* b. 6, l. 12, fols. 232r–234v.

money, giving instructions that it should be bequeathed to his children.[93] The rest were always debtors, who, upon the approach of their death, beseeched their creditors to pardon them, avowing their wretched financial condition, and so gave an indication of their rank on the social scale. Some (seven out of thirty-one) did not draw up a will, since they had nothing to bequeath, and mentioned no relative. They asked for forgiveness in the words recorded by the comfort givers of the Confraternity of San Giovanni Decollato, this being the last opportunity granted to them to leave some remembrance of themselves on earth.

Surprisingly, it is impossible to find any sort of connection between dissection and the type of trespass committed by the person destined for it: those who were dissected were guilty of such ordinary crimes as theft and homicide. Only one had been accused of sodomy.[94] It seemed sufficient that the condemned person should have committed an "atrocious" crime (as contemporary law defined it), that he or she had been executed between January and March, that the *protomedicus* had submitted his request to the governor, and that the cadaver corresponded to the criteria cited above. The case of Alessandro da Spoleto, sentenced for theft and hung on January 21, 1561, in the company of Antonio di Romualdo of Fabriano is a typical example. The latter had been executed for a triple homicide and theft.[95] If the selection of bodies to be consigned to the *Sapienza* had been based on the gravity of the crime, Antonio's cadaver would surely have been fated to serve science. But his mother and two brothers were in Rome.[96] They might have protested such an atrocity being perpetrated on the body of their loved one with the approval of the city's highest judicial authorities. Alessandro, a simple thief, but without

93. Ibid, b. 5, l. 10, fols. 3v–4r.

94. I have succeeded in collecting the following information about the crimes from the extremely fragmentary documentation: Caterina di Lorenzo Corso, executed February 14, 1516, sentenced for having "morto un povero huomo" (ASR, *San Giovanni Decollato*, b. I, l. 3, fol. 32v); Frolio D'Alí, hanged January 13, 1561, and Alessandro di Piero Santo of Spoleto, hanged January 20, 1561, both for theft (ASR, *Tribunale del Governatore, Registrazione d'atti*, l. 38, fols. 145r, 152v; Bernardino da Treviso, hanged January 9, 1573, for theft and other crimes (ibid, l. 62, fol. 135r); Silvestro Pietrobello of Montereale, hanged December 18, 1573, for diverse thefts (ibid, l. 63, fol. 228r); Tommaso da Firenze, nicknamed "Malinconia," hanged January 15, 1574, and Salvatore di Sessa for diverse thefts on February 5, 1574 (ibid, l. 65, fols. 43r, 88v); Annibale Furlano of Parma, hanged January 14, 1575, for theft and sodomy, and Paolo Buscatti, hanged January 28, 1575, for homicide (ibid, l. 66, fols. 18r, 39r); Stefano di Galeazzo of Bracciano, hanged on January 10 or 11, 1578, for theft and homicide (ibid, l. 74, fol. 55r); Antonio Bergamasco, hanged on February 6, 1579, for theft (ibid, l. 75, under the date February 6, 1579).

95. ASR, *Tribunale del Governatore, Registrazione d'atti*, l. 58, fol. 157v.

96. They are quoted in the testament. ASR, *San Giovanni Decollato*, b. 2, p. 4ff., 162vff.

relatives in Rome, fulfilled the requirements of public order while more effectively safeguarding the reputations of the physicians and of the governor of Rome. Alessandro Benedetti was unambiguous on the criteria: "Only the ignoble, the unknown from foreign lands, can be solicited for dissection, therefore, without giving affront to the community and without the next-of-kin disputing their rights."[97]

This recourse to the cadavers of outsiders, which was intended to avoid giving offense to the local population, was motivated by a desire on the part of the civic and academic authorities to ensure that the entire process of the anatomy—from execution to burial—should be carried out with the least disturbance and without the confusion or tumult that the relatives or friends of the deceased could provoke.[98] People were bound to resent the fact that a loved one had been subjected to an operation that bordered on the sacrilegious.

Among the criteria governing the selection of the cadaver, the most significant, perhaps, was that the subject had to be low-born. This was in addition to the stipulation that the person designated for dissection should be someone foreign to the territory and the community (*ignoti*) where the condemnation and dissection took place. Benedetti has remarked on this, and it is implicit in the fact that most of the dissected bodies were victims of hanging, a form of execution usually reserved for the lower classes, which carried with it a mark of infamy. *Ignobilitas* is the term that described the figure of the dissectible body. For the anatomist, dissecting a corpse that was considered ignoble created a path through the shifting limits of what was tolerable, and freedom from both ecclesiastical censure and from public condemnation for the victim and all those who participated actively or passively in the dissection.

AROUND THE CADAVER: BEFORE AND AFTER THE ANATOMY

" . . . y remarqua ce qu'il a dict ailleurs combien le peuple s'effraïe des rigurs qui s'exercent sur les cors mort."

M. Montaigne

I have discussed in previous sections the figure of the anatomist and the criteria for choosing the cadavers to be dissected, while considering at the

97. Benedetti, *Anatomice,* l. I, fol. AIII.

98. I have never come across any mention of the sort of popular protests against anatomists and surgeons occurring in England or France in the eighteenth century taking place in Rome: see P. Linebaugh, "The Tyburn Riot against the Surgeons," in *Albion's Fatal Tree,* ed. D. Hay (London, 1975), pp. 65–117. G. Ferrari ("Public Anatomy Lessons," p. 88) reports an incident occurring in Bologna in 1681, taken from A. F. Ghiselli, *Memorie antiche manoscritte,* preserved in the Biblioteca Universitaria, Bologna (MS. 770).

same time the statutory norms that governed the relationship between them and introducing what will now become my focus here, namely the journey of the victim from prison to burial. Such a reconstruction has been made possible by the discovery of an archival repository hitherto untouched in previous studies of the subject: the records contained in the *Arciconfraternita di San Giovanni Decollato* in Rome.[99] This confraternity was one of those pious sodalities that came into being in great numbers at the end of the fifteenth century, drawing spiritual and religious inspiration from the *devotio moderna*.[100] The movement aspired through such forms of association and the practice of charity in daily life to revitalize Christianity, which was seen as having been corrupted by human appetites and as being rooted at that point in a religiosity that had become almost exclusively formal. The *Arciconfraternita* had been founded in 1488 by a group of Florentine aristocrats residing in Rome, on the model of the *Compagnia di Santa Maria della Croce al Tempio* established in Florence in 1343.[101] They directed their charitable work toward providing comfort and moral and material assistance to persons condemned to death.[102] Similar groups came into being all over Italy between the end of the fifteenth century and the first half of the sixteenth, with a few surviving until the nineteenth.[103]

The brethren of the *Compagnia della Misericordia* visited prisoners and attempted to convince them of the justice of their sentence. They tried to lead them to repent of their sins, and to comfort them with the hope that they would find salvation in the hereafter. The confraternity had the

---

99. See note 91 above. On the gap in the records, see L. Firpo, "Esecuzioni capitali in Roma (1567–1671)," in *Eresia e Riforma nell'Italia del Cinquecento* (Florence and Chicago, 1974), especially pp. 313–15.

100. On the *devotio moderna* and for a bibliography on the subject, see H. Jedin, *Handbuch der Kirchengeschichte* (Freiburg and Basel, 1985).

101. In Florence the *Compagnia della Misericordia* worked toward the same ends. See, E. Capelli, *La Compagnia del Neri: L'Arciconfraternita dei Battuti di S.ta Maria della Croce al Tempio* (Florence, 1927), and, more recently, S. Y. Edgerton Jr., *Pictures and Punishment: Art and Criminal Prosecution during the Florentine Renaissance* (Ithaca and London, 1985), especially chapters 4 and 5.

102. On the Archconfraternity of San Giovanni Decollato, see V. Paglia, *La morte confortata: Riti della paura e mentalità religiosa a Roma nell'età moderna* (Rome, 1982), and I. Polverini Fosi, "Pietà, devozione e politica: Due confraternite fiorentine nella Roma del Rinascimento," *Archivio Storico Italiano* 149 (1991): 119–61. On these groups in general, see A. Prosperi, "Il sangue e l'anima: Ricerche sulle compagnie di giustizia in Italia," *Quaderni storici* 15, no. 51 (1982): 959–99.

103. The first sodality dedicated to comforting those condemned to death appears to have been the *Compagnia di Santa Maria della Morte* founded in Bologna in 1336. For Padua, see G. De Sandre Gasparini, "La confraternita di San Giovanni Evangelista della Morte in Padova e una 'riforma' ispirata dal vescovo Barozzi," in *Miscellanea G. G. Meersseman* (Padua, 1970), 2:765–815.

prerogative, moreover, of setting free annually (and thus rehabilitating) one of the condemned persons languishing in the papal prisons. The books kept by its steward (*provveditore*) not only contained the society's decrees, the listing of expenses, the dispensation of alms, the annual and semiannual renewal of positions, and the election of new members, but also recorded brief accounts of all the death sentences carried out in Rome. These books thus constitute a primary source for the history of criminal justice in the city. Of relevance to this study is the fact that they regularly mention the temporary transfer of the bodies of the executed criminals to the physicians of the *Sapienza* "for the purpose of doing an anatomy." From the examination of these registers for the sixteenth century it can be established that there were at least thirty-one dissections performed on condemned criminals.

The guidelines for the dispensing of compassion and the duties of the good comforter are described in detail in those manuals compiled by members of the confraternity for their brethren, especially the most recently initiated. Pompeo Serni's *Trattato utilissimo per confortare i condannati a morte per via di giustizia*,[104] for example, in addition to setting forth the general rules to be followed for a successful mission, also describes every possible eventuality that the comforter might encounter, and gives relevant advice. These manuals never mention the fact that a condemned person might later be dissected. Before the execution actually took place neither the victim nor the comforters could know who would be turned over to the doctors and, thus, no special behavior was called for. But the condemned destined for execution seemed always to live in fear that their bodies "would be given to physicians to do an anatomy," and this provoked great anguish. They believed that a body thus profaned and disfigured could not hope to be recomposed on the day of the final resurrection, and of course the prospect of one's body being cut up could not have been pleasant for anyone to contemplate.[105]

The brothers of *San Giovanni Decollato,* during the afternoon, would be informed in writing by the proper judicial authority (sometimes the

---

104. BAV, MS. Vat. Lat. 13596, written in 1665. Preceding it, but equally important, are T. Crispoldi, *Alcune ragioni da confortare coloro che per la giustizia publica si trovano condannati a morte* (Ancona, 1572), and Z. Medici, *Trattato utilissimo di conforto de' condannati a morte per via di giustizia. . . .* (Ancona, 1572). To these should be added G. B. Scanarolo, *De visitatione carceratorum libri tres* (Rome, 1655), even though the act of comforting constitutes only one of the many charitable practices described.

105. G. Manara, *Notti malinconiche. Nelle quali con occasione di assister' à condannati a morte, si propongono varie difficoltà spettanti a simile materia. Serviranno per istruttione à confessori, confortatori, e altri assistenti nelle conforterie* (Bologna, 1668), p. 295 (cited in P. Camporesi, *La casa dell'eternità* [Milan, 1987], p. 173).

senator, but usually the governor of Rome) that a condemned man would be executed the next day. The *provveditore* of the confraternity, after reading the communication and satisfying himself that it was genuine, would select two comforters from the group: one of them would carry a tablet with the image of the crucified Christ, the other a book of prayers.[106] The two, along with other members,[107] would go that night to the prison where the condemned person was held, hoping to persuade him to repent and make contrition,[108] to hear his confession and to receive his last testament. They would assemble first in that place designated for the dispensation of comfort. The intended victim would then go alone with the confessor to an adjoining chapel and, finally, to the testamentary room where, with the assistance of the comforters, he wrote or dictated his last wishes and, sometimes, a letter to his family.[109] Then the chaplain celebrated Mass, during which the condemned person received communion "with great devotion."

At this point the victim was ready for his punishment. In the early morning hours the executioner would appear and tie a rope about his neck, while the brothers of the confraternity chanted litanies and one of them placed before his face the tablet with the image of the crucified Christ. They would leave the prison in this manner, joined by other members of the association. The procession was led by the sacristan and the brethren carrying the cross and torches, followed by the condemned person and the two comforters seated together on a cart pulled by a pair of horses, and trailed by guards. The cortege was usually mobbed by curiosity seekers when it arrived at the place of execution; the square by the bridge of Sant'Angelo (Piazza di Ponte) was the usual venue for hangings. The brothers went on singing and the victim gazed at the tablet that the comforter held before him. The condemned person then went to receive absolution and to recite his last prayers in the chapel belonging to the

---

106. ASR, *San Giovanni Decollato*, l. 19, fol. 182r, 22v. These documents are also cited in Paglia, *La morte confortata*, pp. 115 ff.

107. Among whom one always finds the *provveditore*, the chaplain, two sacristans and the factor (ASR, *San Giovanni Decollato*, l. 19, fol. 22v), and Paglia, *La morte confortata*, p. 116.

108. I am not going to pause over this point, even though I consider it of great interest. The techniques of persuasion used by the comforters when they encountered obstinate cases merits a separate study. The phenomenon of impenitent offenders and the consequent attitudes adopted by the comforters to reduce them to a recognition of their sins and to repentance are illustrated with a wealth of detail in the treatise by Pompeo Serni and in those by Tullio Crispoldi and Zenobio Medici, cited previously.

109. The testaments of persons condemned to death are transcribed either in the books of the *provveditore* (ASR, *San Giovanni Decollato*, b. 1–8) or in volumes specifically designated for them (ibid, b. 16, ll, 33–34: *Testamenti 1565–82*).

confraternity of San Giovanni Decollato adjacent to the bridge. Finally, with his back to the platform, he retreated toward the scaffold: only the executioner, a comforter, and the image of Christ accompanied him to the gibbet (fig. 31).[110]

At the top of the ladder the victim was dropped into space by the executioner, who pushed down on his shoulders while another official pulled on his legs, causing an almost instantaneous death. At this point the crowd of spectators and the other players in the ceremony left the site of the execution, and the body of the victim remained publicly exposed, as a warning and example, until evening.[111] Some of the brethren of San Giovanni Decollato, known as the "Thirty of the Evening,"[112] gathered at about nine o'clock in the church of Sant'Orsola and returned in procession to the bridge of Sant'Angelo to lower the body and load it on a bier carried by two porters for burial in the church of the confraternity. A few

---

110. M. Montaigne, in his *Journal de voyage en Italie, la Suisse et l'Allemagne 1580–81* (I have used the Paris: Les Belles Lettres, 1946 edition) relates, in a page recorded by his secretary:

> L'onsième de janvier [1581], au matin, comme M. de Montaigne sortoit du logis à cheval pour aller in Banchi, il rancontra qu'on sortoit de prison Catena . . . Il s'arresta pour voir ce spectacle. Outre la forme de France, ils font marcher devant le criminal un grand crucifix couvert d'un rideau noir, et à pied un grand nobre d'homes vetus et masqués de toile qu'on dit estre des jantils homes et autres apparans de Rome, qui se vouent à ce service de accompaigner les criminels, qu'on mene au supplice et les cors des trespassés, et en font une confrerie. Il y en a deus de ceus là, au moines, ainsi vetus et couvers, qui assistent le criminel sur la charette et le preschent, et l'un d'eus lui presante continuellement sur le visage et lui faict baiser sans cesse un tableau où est l'Image de Nostre Seigneur. Cela faict que on ne puisse pas voir le visage du criminel par la rue. A la potance, qui est une poutre entre deus appuis, on lui tenoit tous-jours cete image contre le visage, jusques à ce qu'il fut élancé. (Pp. 108–9)

111. Concerning the modes of execution, see the many accounts of capital punishment preserved in ASV, BAV, and BN cited by Paglia, *La morte confortata*, p. 106, note 32; A. Ademollo, *Le annotazioni di Mastro Titta, carnefice romano* (Città di Castello, 1886); A. Keller, ed., *A Hangman's Diary: Being the Journal of Master Franz Schmidt, Public Executioner of Nuremberg, 1573–1617* (London, 1928); and Firpo, "Esecuzioni capitali." I have also been mindful of remarks by Thomas Laqueur in his lecture on public executions and Carnival in England during the seventeenth century, held in Paris at the École des Hautes Études en Sciences Sociales in the seminar directed by J. Revel in 1988. For Florence, see S. Y. Edgerton, *Pictures and Punishment*, pp. 139 ff. There is no discussion in the latter work, in regard to hanging, of pulling down on the victim's legs, nor of the executioner pushing down on his shoulders to shorten the agony: "The condemned person simply swung free of the ladder and perished by strangulation" (p. 142).

112. The *Trenta della sera* came into being by decree of the *provveditore* and were approved by a vote of the entire sodality in July 1511. They were selected by the *provveditore* himself and could be expunged from the books of the confraternity if they failed in their office on three occasions (ASR, *San Giovanni Decollato*, b. I, l. 2, fol. 16v).

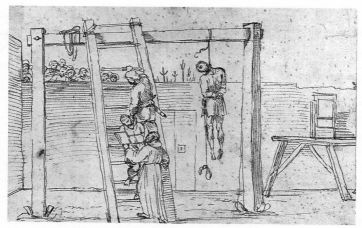

Figure 31   Annibale Carracci, "A Hanging," circa 1599 (Windsor Castle, Royal Library). In the course of the sixteenth century all the bodies dissected in Rome, in accordance with the regulations dictated by the statutes of the College of Physicians, were of criminals who had received capital punishment. Already by the end of the fifteenth century in Rome the Confraternity of San Giovanni Decollato took charge of those sentenced to death from the time of their detention and execution to their burial. It was the duty of the members to comfort the condemned, to seek to bring them to repentance, and to reassure them about the otherworldly fate of their souls as well as about the dignified burials their remains would receive. These condemned persons often feared that they would end up on the dissecting table. In this drawing Annibale Carracci illustrates the moment when one of these unfortunates is being dragged to the scaffold by the executioner. An adept of the confraternity accompanies him, holding a tablet of Christ's image before his eyes.

days later a simple Mass would be celebrated and the victim's clothing sold.[113] Ordinarily more than one execution took place in a single day.

This was the usual procedure. The ritual and institutional practice differed somewhat when the condemned person was to be consigned to the anatomists for dissection.[114] The 1536 statutes of Roman physicians, in force until the 1676 reforms, do not mention how many public anatomies should or could be held in an academic year and, consequently, how many

113. For a description of these practices between 1506 and 1585, see ibid, b. 1–6, which list the executions. Only in a few cases was the body of the condemned person turned over to his family for burial in his parish. They would be responsible for the expenses of the last rites and of the funeral Mass in such a case.

114. There would be discussions in the general assemblies of the confraternity of the procedures to be followed in these cases; on one such occasion, January 30, 1558, the discussion was sufficiently important to warrant an entry in the book of the *provveditore* without further explication of the terms: "M. Betto da Castello Taver. parlocci delle notomie il Governatore et Gio. Goncozzi presono cura d'andare al rettore dello studio e Gio. Goncozzi referí aver avuto il denaro" (ibid, b. 2, l. 4, fol. 39v).

cadavers would be needed.[115] Generally, for the period 1506–85, the available documentation suggests that one dissection was performed annually, although frequently none at all took place.[116] Two were held in 1560–61, 1572–73, 1574–75, and 1592–93; three in 1573–74, all performed by Scipione de' Rossi; and four in 1587–88.

The order signed by the senator or, more frequently, by the governor stating that the cadaver of the condemned person was to be consigned to the physicians or to the students of the *Sapienza* was received by the confraternity the night following the execution. Presumably the judicial authorization would have been requested in advance by the physicians through the rector of the university. Although there is little information on the matter, it seems reasonable to suppose that the supplication from the latter or from the *protomedicus* and the selection of the body for the public anatomy, following the criteria outlined above, preceded the execution, probably by several days. The selection process was clothed in strict secrecy, for obvious reasons. News about the matter might influence the rites of comfort, or, even worse, reach the ears of the victim. The comforters' words hinged essentially on three points to which the condemned person would be especially sensitive, and which furnished a sort of key with which to persuade him into resignation and the acceptance of a violent and premature death at the hands of the judiciary: (1) an assurance that his condemnation would not stain the reputation of his family; (2) the hope that his soul, repentant and contrite, would be given a place in Purgatory or even straightaway in Paradise; (3) the guarantee that his body, safeguarded by the confraternity, would be protected from any possible depredation,[117] and would receive a dignified burial.[118] The absence of any one of these three fundamental requisites could have caused the collapse of the entire psychological structure on which the act of comforting was based.

The governor's decree[119] usually arrived while the "Thirty of the Eve-

115. The statutes of 1676 speak of "cadaver unum vel plura" provided by the governor to the *studium* (*Statuta 1676*, chapter 51: *De anatomia exercenda*).

116. The confraternity did not furnish cadavers for dissection in the years 1506–11, 1513–15, 1517–22, 1558–60, 1562, 1564, 1566–72, 1576–77, and 1580–82. I should like to repeat that it is impossible to establish the frequency of public demonstrations between 1523 and 1555 because of the lack of documentation for these years.

117. See A. Prosperi, "Il sangue," p. 963.

118. In particular, see P. Serni, *Trattato utilissimo per confortare i condannati a morte per via di giustizia* (1665). BAV, Rome, Ms. Vat. Lat. 13596, pt. 2.

119. Only in two cases, those of Caterina di Lorenzo and of Gian Domenico Sforza, was the senator the competent judicial authority to issue the mandate (ASR, *San Giovanni Decollato*, b. I, l. 3, fol. 32v and b. 4, l. 8, fol. 65v). In the case of Camilla Mansueto it was

ning" were gathered at the church of Sant'Orsola preparing for the recovery of the bodies and for their burial.[120] After reading the mandate brought to them by a messenger or by the "ministers of the anatomy" themselves, the brethren would go to the bridge of Sant'Angelo to lower the bodies of the victims. The one destined for dissection could then be consigned directly to the representatives of the university at the place of execution, and the latter would be responsible for bringing it to the *Sapienza*. Alternatively, the body might be laid out on a bier and transported by the confraternity in procession to the university. This route differed only slightly from the one usually taken from the bridge of Sant'Angelo and the church of San Giovanni Decollato.[121] As a rule it was the anatomist himself who went to San Giovanni to take possession of the body, while the brothers were occupied with burying the others executed that day and with administering the funerary rites.[122]

Two points need to be stressed here. First, it was essential that the consignment of the body to the physicians take place out of sight, during the night while the city slept. The same logic dictated that the body should be handed over in the church of San Giovanni Decollato rather than in the square, at the place of execution, where relatives or even the merely curious could have witnessed the consignment.

Second, the brief notes found in the books of the *provveditore* of the confraternity always specified that the body was consigned to the "minister of the anatomy," and the name of the person who had come to claim it was even occasionally supplied. Beginning in 1583 the records mention that the physicians would give a receipt when they took possession of the

the vicar of Rome who turned her body over to Angelo Antonini (ibid, b. 7, l. 14, fols. 99r–v), and for the body of Santi Palovio, also consigned to Antonini, it was Prospero Farinacci who did so in his capacity of "luogo tenente del Monsignor della Camera" (ibid, b. 7, l. 15, fol. 184r–v).

120. On other occasions the message reached the gallows at the moment that the bodies were being loosened from the ropes, or very rarely, in the church of San Giovanni Decollato, a few moments before the body was buried, as occurred in the case of the cadaver of Alessandro di Pietro Santo of Spoleto (who was hanged on January 21, 1561): "Poi a ore 21 in circa raunati i nostri trenta della sera in S.ta Orsola s'andorno a stacare e portati alla nostra chiesa per sepelitura e fu sepelito Ant. E venuto uno ministro dell'anatomia con uno mandato del Governatore di Roma che li si dovessi dare il corpo di Alexandro e così si fecie" (ibid, b. 2, l. 4, fol. 162v).

121. This occurred at least four times: ibid, b. 2, l. 4, fol. 305r and b. 4, l. 9, fols. 64v, 65v and 100v.

122. This was the procedure in all the other documented cases. Only once did the students go to fetch the body: "e così [Giovanni di Natale] fu portato alla nostra compagnia dove si aveva uno mandato del Governatore di Roma a dire predetto giustiziato si consegnassi a cierti studianti alla Sapienza" (ibid, b. 2, l. 4, fol. 29v).

body,[123] though it remained anonymous between 1583 and 1586. This happened after a period of three years from 1580 to 1582, during which no bodies had been turned over to the anatomists by the confraternity. There is no documentation available to explain this lapse, but the change in the modalities of registration (which provided for the anonymity of the dissected body and included the request for a receipt) immediately after this suggest some sort of consequential link, if not an actual conflict, between the parties involved. I can only present hypotheses to explain these altered procedures. The custom of issuing a receipt to the brethren of San Giovanni perhaps preceded its registration in the books of the *provveditore*. This could be significant: with the receipt, the confraternity documented the loan of the cadaver to the physicians of the *Sapienza* and guaranteed the legality of the transaction; at the same time it delegated responsibility for the care of the cadaver to the physicians during the period of time in which they would be using it. It is clear that the purpose of the bureaucratic system that had been set up was to scrupulously keep track of the cadaver, following it every step of the way, establishing a rigorous chain of responsibility: from the guards, to the confraternity, to the *Studium Urbis*. The entire operation proceeded under the aegis of the judicial authorities, which had absolute jurisdiction over the body of the condemned;[124] the powers of the confraternity and of the university emanated from it, but in a temporally limited way.

The anonymity of the body obviously related to a different problem and had a twofold purpose. On the one hand, it worked to keep a veil of secrecy over the ignominious fate of the executed person and, on the other, it shielded the confraternity, to whom the body of the condemned had been entrusted with a recommendation to preserve it from any type of profanation, from the possible protests of the family and friends of the deceased. In addition, anonymity, along with the requirement of the receipt, were additional precautionary measures employed to preserve the anatomical procedure, the cadaver, and all the concerned parties, from the conflicts and remonstrations of a moral, religious, and even emotional

123. For example, on January 28, 1583, "Venne un ordine di Mons. Gov. che la compagnia consegnassi uno delli detti corpi a la Academia de Notomisti; fu consegnato et ne fecero ricevuta" (ibid, b. 6, l. 12, fol. 90r). All the subsequent entries are similarly couched.

124. This is sustained also by A. Prosperi ("Il sangue," p. 992): "Del resto, che il corpo del condannato fosse materiale disponibile per ogni tipo di esperimenti già prima dell'esecuzione è provato dal fatto che un sospettoso granduca fiorentino ne fece uso per studiare gli effetti di veleni ed antidoti (ASF, *Strozziane*, ser. I, 97, fols. 1r–7v)." See also A. Andreozzi, "La vivisezione anatomica dei condannati a morte sotto Cosimo I duca di Toscana," *Rivista di Discipline Carcerarie* 32 (1907): 27–33.

nature, which might have occurred during the years in which the practice was trying to construct its own legitimacy.[125]

The physicians kept the body of the condemned while it was being considered for dissection for a varying number of days, normally from five to ten. Only in three cases was it retained longer, even as much as three weeks, as on one occasion when a second dissection was scheduled at short notice.[126] The delay in restituting the body was certainly not dictated by scientific reasons but arose simply from the wish to minimize the additional expenses of transport.[127] The prolonged retention of the cadavers at the *Sapienza* is evidence of the fact that there was a sort of programming of dissections that also affected the selection of the bodies. It was therefore not just a matter of improvisation simply dictated by the occasional request of the university's physicians to the governor of Rome, who at the last moment would hurriedly try to locate a condemned person whose attributes corresponded to the needs of the doctors and prevailing conventions. The doctors and the governor were certainly aware of such programming, but so were the members of the confraternity, from

125. The case of the Roman Paolo Buscatti, who was executed and dissected in 1575, is an example of this prudent policy. It had been long maintained that the criteria behind the selection of the cadaver dictated that it should not be a Roman, precisely so as to avoid protest, tumult, or confusion, and that Buscatti was one of the two Romans who, according to the documentation, was destined for the dissector's knife. His execution took place on January 28, and normally his body should have been buried on the evening of the 29th. At least this is what the people understood on the basis of custom. The next day, the 30th, "fu consegnata la testa di Paolo Buscatti giustiziato a notomisti e per loro a m. Antonio Rosena notomista per vigore di uno mandato del Governatore di Roma" (ASR, *San Giovanni Decollato*, b. 5, l. 10, fol. 7r). At this point, no one could have known or suspected that such an occurrence had taken place: without even considering that the victim had been hanged and not decapitated we are left with the mystery of how the head had become separated from the neck.

126. This happened in the case of the body of Bernardino di Giovanni of Treviso, executed on January 9, 1573, and returned to the confraternity of San Giovanni Decollato on the 23rd of that same month, the date on which the body of Gian Domenico Sforza was ceded to the physicians (ibid, b. 4, l. 9, fols. 65r–v). It also occurred for the body of Tommaso di Raffaele of Florence, executed January 15, 1574, and returned on a day coinciding with the capital punishment of Salvatore di Sessa on February 5 (ibid, fol. 102v); and, finally, for the body of Annibale Furlano, hanged January 14, 1575, and returned on February 2 together with the head of Paolo Buscatti, which had been consigned to the anatomists on January 30 (ibid, b. 5, l. 10, fol. 8v). Much earlier, in 1561, the body of Frolio d'Alí was returned to the confraternity for burial after eight days, while the cadaver of Alessandro di Pietro Santo was left to the physicians of the *Sapienza* (ibid, b. 2, l. 4, fol. 162v and r).

127. The expenses borne by the confraternity to pay the porters who transported the cadavers are entered at the end of each execution. For example, on January 28, 1561, sixty *baiocchi* were expended to carry back the body of Alessandro da Spoleto from the university to the church of the confraternity (ibid, b. 2, l. 4, fol. 163v).

the moment they realized that the body they had previously delivered to the physicians was being kept longer than was customary. Nevertheless, the name of the executed person marked for dissection unquestionably remained secret until the last moment.

During the anatomy every single body part was collected and placed in a basin, which is visible in many iconographic representations of the lesson (figs. 1, 3, 8, 27). At the end of the ceremony the body and all its separate segments were placed in a chest and reconsigned to the confraternity for burial.[128] In the statutes of the Roman physicians from 1536 until 1584, at rubric seven in the chapter dedicated to public anatomies, it is stated that "someone should be made ready who can transport the parts to the cemetery." The actual procedure can be traced through a case reported in unusual detail in the books of the confraternity of San Giovanni Decollato:

> Friday, 5 January [actually February] 1557. That morning we were informed by a messenger from the physician M. Rialdo [Colombo] that we should go that night at ten o'clock to retrieve Giovanni di Natale of Forlí who had been executed and whose body had been dissected. Then the Thirty of the Night were informed and we gathered in Santo Eustachio and at the usual hour an emissary came from his lordship to say that the students were ready. We went to fetch him at the *Sapienza* and carried him to our quarters and many students accompanied him to keep him company and they gave us two new torches and to each of the brothers two candles to hold in their hands, and then they gave to me, the *provveditore*, b. 85 to pay the porters, and when they [the students] left our Company they profusely thanked all the brothers.[129]

The practice described here was followed, in its essential features, in all the other known cases. From the evidence furnished by the book of the *provveditore* it would appear, then, that it was not the College of Physicians that concerned itself with the removal of the remains and their burial, as was provided for in the statutes: a group of students, and rarely the physician instructors themselves, accompanied the cadaver, which was transported by the porters paid by the confraternity along with the Thirty of the Evening, as far as the church of San Giovanni Decollato. Here the body received an honorable burial, concluded by and at the expense of the confraternity.

128. On February 3, 1573, the Confraternity of San Giovanni spent thirty *baiocchi* "per una cassa che si comperò per mettervi detto corpo [of Domenico di Mariano Sforza]" (ibid, b. 4, l. 8, fol. 67r).

129. Ibid, b. 2, l. 4, fol. 29r.

MASSES AND ALMS: DISSECTION AND THE AFTERLIFE

Certain discrepancies exist between the provisions of the statutes and the practice effectively followed in the disposal of the remains of the executed dissected person. As stated earlier, the eighth rubric of the 1536 statutes established that, at the completion of the anatomy, funerary rites should be performed and at least twenty Masses celebrated for the soul of the condemned person. In the book of the *provveditore* the matter of supplementary Masses is only recorded twice and only because of the special status of the deceased: in 1512 for the funeral rites of Giovanni da Monte nine Masses were celebrated,[130] and in 1563 Tommè di Bonanno "was accompanied by the Thirty as usual and Petro Coli our chaplain, who together performed four simple Masses in our church and one sung Mass, as well as the offices of the dead for his soul."[131] Nothing is specified in other cases. Though the statutes established a special provision for the dissected victims of execution, it would appear that it was hardly ever followed.

It may be useful to examine the significance of the injunction for twenty supplementary Masses for the soul of the dissected body, since the funerary rites for a normal person condemned to death called for only one Mass. A digression on Masses for the dead and on the theological principles that animated religious beliefs in those years can provide some additional information on the somewhat special status of the dissected cadaver.

As has been noted, comforters would resort to any means whatever to induce a condemned person to repent of past sins. In fact, through the confession that he was obliged to make and through his consequent contrition, the condemned could hope to be pardoned anew by a heavenly court and to be admitted before God, either immediately or following a sojourn in Purgatory. Even the most heinous sins, those that would induce an earthly court to promulgate a death sentence, could be forgiven to the extent that a profoundly repentant condemned person could even be admitted directly into God's presence in Paradise: "And so fervently could the sentence of death and punishment be accepted, that the gallows and execution block would seem to him like a holy martyrdom since at once without Purgatory he would fly up to the heavens," as Zenobio Medici states in his manual for the comforters of the condemned.[132] It can be

---

130. Ibid, b. 1, l. 2, fol. 31v.
131. Ibid, b. 2, l. 4, fol. 222v.
132. *Trattato utilissimo*, fol. 10r. In Tullio Crispoldi's manual for comforters we read once more the theme of divine pardon: " . . . di quelle tante minaccie della scrittura vecchia, la maggior parte, s'intendono delle pene temporali e non delle eterne, imperoché qual hora il peccatore si converte, o per sua volontà, o per qualche necessità, il Signor Dio non gli

argued from this text that capital punishment not only promises pardon, but also can bring about the possible remission of divine punishment for the sins committed that could have been exacted by the torments of Purgatory—or at least for a good part of them. Bartolomeo d'Angelo, addressing himself to a person facing execution, wrote: "You are shown that the blessed God loves you since He permits you to die executed (*giustiziato*). What does it mean to be executed, if not to be justified; what does it mean to die justified if not to pay with one's life that debt that the soul owes to the world and to the devil for the sins committed, so that it may remain free and return to God its creator?"[133]

In conclusion, the soul of the repentant and contrite person condemned to death went to Paradise or, in the worst of cases, to Purgatory: everything depended on whether he or she had time on earth to expiate his or her transgression, whatever it might have been; Hell was only for someone who persisted in sin, either as an unbeliever or a heretic. This was a conception of divine justice, then, that left to the dying and to those condemned to death ample hope in the possibility of a total rehabilitation, as in this specific case.[134] All this conjures up an image of a sparsely populated Inferno, a little like Origen had imagined it, inhabited only by heretics and unbelievers. In contrast, Paradise and Purgatory were swarming with the souls of the faithful who, after sinning, had finally repented.[135]

niega mai la sua grazia, anzi gli va incontro, et l'abbraccia, et lo fa rivestire et non gl'imputa che la necessità l'abbia fatto tornare, anzi riprende agramente chi di ciò non s'allegra, dicendo, che in cielo si fa similmente festa d'ogni peccatore che si converte" (*Alcune ragioni di confortare*, fols. 33v–34r).

133. *Ricordo del ben morire* (Brescia, 1589), p. 369. The same tone is taken by Roberto Bellarmino: "La terza classe di uomini si potrebbe chiamare felice se conoscessero il loro vantaggio. Perché o vengono uccisi con giustizia o ingiustamente. Se con giustizia, la loro morte puo giovare loro come soddisfazione davanti a Dio purché detestino seriamente il loro peccato e accolgano volentieri la morte in espiazione delle loro colpe." And he adds: "Se invece vengono uccisi ingiustamente e perdonano il cuore a chi fu causa della loro morte, saranno imitatori del Redentore che pregò il Padre per i suoi crocifissori" (*L'arte di ben morire*, trans. C. Testore (Turin, 1946), pt. 2, chapter 14, p. 418. On the subject, see also A. Tenenti, *Il senso della morte e l'amore della vita nel Rinascimento* (Turin, 1978), pp. 329 and 350, note 164 (1st ed. Turin, 1957).

134. A medieval doctrine of sin that called for greater rigor in expiation after death, especially for the perpetrators of criminal acts, seems to have been definitely set aside. On the question, see J. Le Goff, *La naissance du Purgatoire* (Paris, 1981). Penance and the sacrament of confession attained, at least from the beginning of the fifteenth century, a preeminent role in forecasting the destiny of souls and the design of the geography and the "demography" of the hereafter. See the discussion in T. Tentler, *Sin and Confession on the Eve of the Reformation* (Princeton, 1977), especially the chapters "Sin: The Subject of Confession" and "The Working of the Sacrament of Penance."

135. St. Jerome himself affirmed, although he positioned himself in opposition to Origen: "Qui enim tota mente in Christo confidit, etiam si ut homo lapsus mortuus fuerit in peccato, fide sua vivit in perpetuum" (*PL*, XXIV, col. 704). Cf. J. Le Goff, *Naissance*, p. 90.

It could be supposed, however, that many of those condemned to death who had been tried and punished for atrocious crimes would not have immediate access to Paradise and would have to spend some time in Purgatory serving their punishment, in spite of their executions and in spite of the encouraging words of Zenobio Medici and Robert Bellarmine.[136] No longer able to accumulate indulgences to shorten their painful passage through the flames of Purgatory through their own good works, the deceased could only count on the intercession of the living. This intervention essentially took three forms: prayer, charity, and, especially, Masses.[137]

Those condemned to death were often unable to leave sufficient funds in their wills to pay for very many Masses for the salvation of their own souls, nor did they have relatives or friends disposed to remember them in their prayers or in their pious acts. The Confraternity of San Giovanni Decollato tried to ensure that even the condemned, outcasts of civic and ecclesiastical society whose memory had been suppressed, would, through the recollection and intercession of the living, have efficacious advocacy for their souls, in case they were languishing in Purgatory. A Mass for the souls of all the condemned was solemnly celebrated on the day that the confraternity paid homage to its patron saint, and on that occasion the ropes used for the executions in the course of the year were burnt. Another Mass was celebrated, but this time for the individual vic-

136. In this regard, see Guillaume d'Auvergne, for whom Purgatory was the continuation of earthly penance (*De universo*, especially chapters 60–65, in his *Opera omnia* (Paris, 1674), 1:76–82. Alexander of Hales expresses himself similarly: the fires of Purgatory purge also "a poenis debitis mortalibus nondum sufficienter satisfactis" (*Glossa in IV libros sententiarum Petri Lombardi*, Bibliotheca Franciscana scholastica Medii Aevi, XV (Quaracchi, 1957). For further discussion on this point, see J. Le Goff, *Naissance*, pp. 325–35. On Peter Lombard, penance and Purgatory, cf. T. N. Tentler, *Sin and Confession*, p. 319.

137. St. Augustine, in his opuscule written for Paolino da Nola, *De cura pro mortuis gerenda*, was the first to sustain explicitly and comprehensively the utility of suffrages for the dead in Purgatory in the form of prayers, charity, and Masses. He adds, however: "Non existimemus ad mortuos, pro quibus curam gerimus, pervenire, nisi quod pro eis sive altaris, sive orationum, sive eleemosynarum sacrificiis solemniter supplicamus: quamvis non pro quibus fiunt omnibus prosint, sed iis tantum quibus dum vivunt comparatur ut prosint" (*PL*, XL, col. 609). The three types of suffrages are all discussed by Gregorio Magno (*Dialogi*, ed. U. Moricca [Rome, 1924]; for prayers, see especially l. IV, 42, pp. 297–300 and for Masses l. IV, 57, pp. 315–17) and the authors of the "Supplement" to Aquinas's *Summa* (Thomas of Aquinas, *Summa Theologica* [Rome, 1894]), *Supplementum, Quaestio* LXXI "De suffragiis mortuorum," especially Article IX, "Utrum orationes Ecclesiae, sacrificium altaris, et eleemosynae prosint defunctis"). On these passages by Gregory and Thomas, cf. J. Le Goff, *Naissance*, pp. 113–114, 367–70. Discussing appropriate suffrages toward the dead, Cornelio Musso, in a sermon entitled *Predica dello stato delle anime* pronounced in 1563, asserted: "Possono ben allegerir quelle pene crudeli [del purgatorio] et per grazia di Christo accelerar la liberazione de' defonti i suffragi, l'indulgenze, l'orazioni, le limosine, i digiuni, gli uffici et soprattutto . . . la celebration delle messe" (C. Musso, *Prediche* [Venice, 1573], l. IV, p. 486. Cf. A. Tenenti, *Il senso della morte*, p. 301).

tim, on the day following his death, and this was in addition naturally to the burial rites. The clothing of the deceased was generally sold by the confraternity one week after the execution, and the proceeds were distributed as charity to the poor in the name of the soul of the departed. Torches and candles were then lit and offered as another form of intercession.[138] It is also plausible that the brethren prayed to alleviate the pains of Purgatory for the deceased, although nothing is said of this practice in the books of the *provveditore*.

Obviously, when the soul of the deceased did not require any earthly intercession, prayers, as St. Augustine stated in his *Enchiridion,* served, nonetheless, to console the living.[139] At any rate, whether or not they were efficacious in shaping the ultraterrestrial destiny of the deceased, or whether they simply remained pure acts of faith, prayers brought to the living merits that they would be able to deduct from their own penances on earth and in the afterlife. This calculation becomes significant in the interpretation of the pious and charitable acts ostensibly addressed to the soul of the dissected cadaver.

All these beneficent practices were of advantage to the brothers, who through them won merits for their own souls. Not only the Confraternity of San Giovanni Decollato but a majority of fifteenth- and sixteenth-century lay confraternities defined their activity and based their very reason for being on this penitential marketplace.[140] It was not by chance that they made themselves the interpreters of a model of the afterlife in which Purgatory came to assume a preeminent position, and that a doctrine of sin was adopted that left ample space for the intervention of divine grace and for the charitable intercession of the faithful.

This provides a context by which we may understand the type of succor extended to the thirty-one condemned persons known to have been dissected in Rome in the sixteenth century. None of them died in a state of obstinacy, and none was condemned as a heretic or unbeliever. Accord-

---

138. Eight candles, as an example, were lit for the last rites of Giovanni da Monte Lauro (ASR, *San Giovanni Decollato,* b. 1, l. 1, fol. 31v). The offering of candles or oil was recommended as a suitable suffrage for the dead by St. John Damascene; on the other hand, Thomas of Aquinas considered it on a par with fasting, an external form of *charitas* (*Summa, Supplementum, Quaestio* LXXI, Article IX).

139. Augustine of Hippo, *Enchiridion,* chapters 109–10.

140. In the thirteenth and fourteenth centuries this commerce was administered primarily at the individual level through testaments (see J. Chiffoleau, *La comptabilité de l'au-delà* [Rome, 1980]). Without losing sight of the persistence even in succeeding centuries of this personal dimension, it seems important to underline the heavy participation of confraternities as intermediaries in the relationship between humans and their otherworldly destinies. Cf. P. Aries, *L'homme devant la mort* (Paris, 1977), especially the chapter "Assurances pour l'au-delà."

ingly, their souls could have been received in Paradise or in Purgatory. The crimes for which they were condemned (see appendix) primarily involved theft, for which the existing penitential doctrine did not impose condemnation to eternal damnation unless the trespasses were against clerical or ecclesiastical property: only then would excommunication *latae sententiae* have been automatically applied. Among the cases, the sources record three homicides (Caterina di Lorenzo, Paolo Buscatti, and Stefano di Galeazzo), a sodomite (Annibale Furlano) and two persons condemned for excesses (Bernardino di Giovanni and Silvestro Pietrobello). In the light of St. Paul's Epistle to the Galatians (5: 17–21), there was no way they could escape the infernal fires.[141] Sodomy, especially, as a sin against nature, was a sure passport to Hell.[142] Murderers were also guilty of a crime that moral theology ranked among those that could be punished by eternal damnation.[143] In spite of this, the then current penitential doctrine and the prayers offered suggest that there was still hope of divine pardon for them. Murderers, in contrast to other condemned persons, received special treatment, at least on paper: twenty Masses in their memory, instead of one, and supplementary alms offered by the College of Physicians to Christ's poor "for the souls of the dissected."[144]

These arrangements conceal three different categories of problems connected with the status of the dissected body (and also with that of the dissecting physician) and they were dictated by a number of concerns. First of all, it is clear from a general reading of these norms that there was an attempt to reward the soul that inhabited the body used by the anatomists for the services it had rendered to physicians and to science by shortening its possible sojourn in Purgatory. In addition, it seems plausible to suggest that anatomists might have considered that a disembowled cut-up body, disarranged for a profane purpose (in contrast to the "sacred" practice of plundering saints' bodies and the cult of relics)[145] resulted in a higher dose of infamy, which had to be rectified in the afterlife. As a consequence a more conspicuous intercession on the part of the

141. "Now the works of the flesh are plain: immorality, impurity, licentiousness, idolatry, sorcery, enmity, strife, jealousy, anger, selfishness, dissension, party spirit, envy, drunkenness, carousing, and the like. I warn you, as I warned you before, that those who do such things shall not inherit the kingdom of God" (Galatians, 5: 19–21.)

142. See D. Doherty, "The Sexual Doctrine of Cardinal Cajetan," in *Studien zur Geschichte der katholischen Moraltheologie*, ed. M. Müller (Regensburg, 1966), 12:121–60, and T. N. Tentler, *Sin and Confession*, pp. 140–44.

143. T. N. Tentler, *Sin and Confession*, p. 312.

144. *Statuta 1531*, chapter 64.

145. For a preliminary sketch of the problem, see J. Gelis and O. Redon, eds., *Les miracles miroirs des corps* (Paris: St. Denis, 1983), especially, the essay by M. Bouvier, "De l'incorruptibilité des corps saints," pp. 193–218.

living was required, as if the body subjected to anatomical inspection had caused the soul inhabiting it an even greater distress than its already sorrowful condition in afterlife warranted. The same logic could be applied to the fact that the dissected cadaver remained unburied after death for weeks at a time, which caused its condition to deteriorate further. It seems highly likely that such reflections were current, if we consider the fact that the statutes of the College of Physicians in Rome recommended that the anatomized body receive prompt burial.

One can however consider the complex of pious practices contained in the statutes of the physicians in a totally different light. The activity of the confraternity involved, on the one hand, charitable work for the relief of the condemned and for their salvation in the afterlife, and on the other, the acquisition, through this same activity, of indulgences for the safeguarding of the brothers' own souls. Similarly, the physicians interceded "for the souls of the dissected," but were also concerned about their own destinies. The participation of physicians and students in the processions that accompanied the dissected bodies to their burial in the *loggia* of the church of San Giovanni Decollato, the payment for Masses by the College of Physicians, the alms dispensed by the college to the poor for the souls of the dissected, which were collected from the fees paid to attend the public anatomy, and the donation to the confraternity of candles and torches, can be interpreted as actions contributing to the rehabilitation of the deceased and, at least formally, also to the professional and moral standing of the anatomist. This interpretation implies that the violation of a body, and with greater reason, of a cadaver, was certainly not an operation without religious and cultural consequences (at least in the way it would be perceived outside the narrow circle of those who performed it). The acts of intercession performed by the physician-anatomists and specified in the university statutes thus assume the character of rites intended to absolve both the physician and the dissected subject from possible attributions of impiety and transgression resulting from certain acts carried out during the dissection.

How are the charitable actions contained in the statutes of the college reflected in the surviving documents? Supplementary Masses, but certainly not as many as twenty, were celebrated in only two of the cases. As for offerings in favor of the dissected cadaver made by anatomists or by students, only two are recorded before 1600: thirty-four *baiocchi* were paid out as alms by Bartolomeo Eustachio for Alessandro da Spoleto;[146]

146. January 28, 1561; cf. ASR, *San Giovanni Decollato,* b. 2, l. 4, fol. 163v.

two *scudi* and fifty *baiocchi* by Pier Matteo Pini for Antonio Blanco.[147] Four other instances are recorded of offerings made by anatomists or students, but they are always in the form of reimbursement for the expenses of transport or burial sustained by the confraternity,[148] and should not be counted as charity. Moreover, after 1573 there is no record even of the participation of anatomists or students in the procession of the dissected cadaver and the rites for the dead.

These lapses could be ascribed to the negligence of the recording party. This hypothesis seems unlikely to me, however, at least as far as the distribution of funds for charity and reimbursements are concerned: every other item of income and expenditure is noted with scrupulous care in the book of the *provveditore*. Even the Masses for the dead, normally recited by the chaplain of San Giovanni, if actually celebrated or paid for by the anatomists, would certainly have been entered in these detailed records. It would therefore appear from these accounts that the conventions covering charitable practices on behalf of the souls of the dissected were not actually applied except in a few cases, and that these now become exceptional. After 1573 there is no documentary evidence of any preoccupation with the fate of the deceased in the afterlife.

Between Saying and Doing

The statutes drawn up by the College of Physicians and then endorsed and ratified by religious and political authorities provide the necessary framework within which the physician, and the anatomist in particular, was to act. In Rome especially, where the statutes bore the signature of the pontiff, they had to satisfy two contrasting demands: that of science, which required direct knowledge of the human body, and that of religion that sought respect toward the dead, their remembrance, and their bodies.

Many of the anatomist's acts, such as the opening and manipulation of the body, or the delaying of burial, were religiously and anthropologically risky. Only a complex system of regulations, strategies, and controls could protect the act of dissection from being sacrilegious and could pro-

147. January 17, 1565; see ibid, b. 2, l. 4, fol. 305r.
148. Of the four, one case seems to me worthy of special notice, if only because of the sum paid by the physicians. The body in question belonged to Tommè Bonanno and had been requested by Pier Matteo Pini, who promised to return it the next morning and to pay all burial expenses (ibid, b. 2, l. 4, fol. 222r). Actually, Tommè's remains were restored to the confraternity only on January 18, but, according to the records, "li anatomisti dettano a noi scudi 3 per queste spese e vennaro sino a nostra chiesa accompagnarlo." The figure was much smaller in other cases and did not cover even half of the transportation expenses.

vide anatomy with an appropriate foundation for a legitimate future without offending any norms of behavior. But, as I have suggested, the conventions regulating the only occasion in which dissection was feasible, the public anatomy lesson, were frequently not respected. Certain controls were evaded and strategies set aside; anatomists did not concern themselves, as they should have, with the burial of the dissected body; alms were not distributed; the twenty supplementary Masses were not celebrated; the cadaver was frequently kept at the *Sapienza* for an extensive length of time and returned only after it was requested by the Confraternity of San Giovanni Decollato or, even worse, was never given back.[149] The failure, especially noticeable after the mid-1570s, to follow these conventions that safeguarded the afterlife of the cadaver while purifying the anatomist, is evidence of the latter's attitude toward his own sphere of activity and its earthly and eschatological consequences.

The explicit and implicit conventions of the statutes and the actions that have been described in this chapter constitute a formal system that was required for the legitimization of the practice of dissection within a cultural context where the religious component was firmly binding. They offered to public opinion an acceptable representation of an activity full of disturbing connotations that fluctuated constantly between the licit and illicit. In this formal apparatus the anatomists had a code of conduct and a means with which to rehabilitate themselves, and the anatomized subject as well. But it remained exclusively a formal apparatus. The governing rules were merely a projection of what the anatomist himself felt to be his formal responsibility within a system of cultural values. On the other hand, the anatomists' application or nonapplication of the normative rules provided the measure of their actual adherence to the culture that had dictated those rules. The original purpose for which these regulations had come into being—knowledge of the human body as the foundation for all medical practice—now began to be seen as a sufficient guarantee of the morality and lawfulness of their work. Proof of this lies in all the dissections that anatomists and students performed without permis-

---

149. In the records of the *provveditore* of San Giovanni Decollato occasionally there is no mention of the dissected body having been returned. Usually, its return is punctiliously noted. Apropos the concern for the restitution of the body demonstrated by the confraternity, the minutes of its assembly held on January 18, 1587, state, "[D]al nostro Governatore della Compagnia fu ordinato al Provveditore che havesse cura che al tempo esso corpo ritornasse alla Compagnia per dargli sepoltura e che si pigliassi cura di farne tornare altri se fussino ancor fuora per simil caso" (ibid, b. 7, l. 14, fol. 54v). It would thus appear that occasional negligence on the part of both anatomists and confraternity members had previously occurred.

sion of any kind. They evaded institutional control, dug up cadavers, and opened them up in secrecy in their own homes.[150]

If dissection was, in the context of a profoundly Christian civilization, a practice that flew in the face of certain fundamental principles of sixteenth-century culture and religiosity, it was nevertheless able to evolve thanks to the establishment of a formal normative apparatus and to the mediation of judicial, political, and religious institutions. However, neither the College of Physicians, the university, the Archconfraternity of San Giovanni Decollato, nor any other authority more or less involved in the organization of dissection reacted to any lapses, even though they must have been aware of them. They never lamented any deficiency either of form or of content in those activities connected to medical research and teaching. A certain spirit of toleration, noticeable by the 1570s, seemed to inform the conduct of the institutions that gradually granted a larger margin of freedom to anatomists. Parallel to this *de facto* emancipation of anatomy from statutory restrictions, which remained in certain respects affirmations of principle, the teaching of anatomy in the *Studium Urbis* acquired, symptomatically, an autonomous chair of its own.

This is not intended to suggest that dissections were carried out in complete freedom. The explicit and implicit precautions regulating the selection of cadavers that remained vigorously in force throughout the century are an example of the constant attention to the cultural and religious significance of the process. An essential and never neglected requirement regarding cadavers acquired through official channels was that they come from the ranks of those sentenced to death, suggesting a consistent association between punishment and dissection. Dissection was conceived as further retribution to be inflicted on the condemned, part of the penalty that would have to be expiated during his or her eventual passage through Purgatory.

In fact, the College of Physicians, by imposing a rather rigorous discipline in the selection of the cadaver, and especially the Confraternity of San Giovanni Decollato, through the doctrine of sin that it had adopted and promoted, were making every effort to guide the image of the dissected person into a regulated environment that would not impose further infamy on the soul of the deceased because of what had happened to the body. This had a twofold purpose: to preserve the memory of the dissected person and to prevent anatomy from becoming an infamous practice in the view of the public.

150. I shall return to this point in chapter 4.

And yet dissection, along with capital punishment, did not escape the stigma of infamy. One need only consider how much the dissected cadavers resembled, in their condition and in the use that had been made of them, the bodies of those who had been condemned to death. In addition to capital punishment, these bodies were often subjected to such punitive action *post mortem* as drawing and quartering. The four sections into which the body was cut up by the executioner were exhibited for several days and then "transported outside" (beyond the city walls) by "masters of justice," at which point they were strewn unburied in the countryside. Treatment of this sort was meted out to obstinate criminals and to those guilty of especially heinous crimes, and to some Jews, unbelievers, and heretics whose bodies had not been burned at the stake. Quartering was a more severe punishment, compared to hanging and beheading, and implied greater culpability. Hope of salvation for guilty souls was greatly reduced in the cases of quartered persons because of the opening of the body and the denial of burial. These are the very features that suggest similarity to the dissected body: quartered corpses, like the dissected, had been profaned and subjected to a series of acts that altered their unitary structure; in addition, in both cases, the bodies were exposed in public for lengthy periods and were left unburied (though for dissected cadavers, burial was only delayed). Desecration, exhibition of the body's internal organs, and delayed or denied burial associate dissection with quartering.

This linkage, which may seem somewhat forced, is clearly displayed in a case involving a death sentence recorded in the books of the *provveditore* of San Giovanni Decollato (January 14, 1587). Rodolfo di Bernabeo of Rome had been hanged along with three companions. After death the latter were quartered "and the pieces carried outside Rome as an example." On the other hand, "Rodolfo di Bernabeo was not quartered but by order of Mons. the Governor of Rome was given for anatomy to the students of the *Sapienza*."[151] In this case, dissection took the place of

151. ASR, *San Giovanni Decollato*, b. 7, l. 14, fol. 54r. There is an entry at fol. 56v, with the date of February 15, that employs a formula that is as unusual as it is macabre concerning the restitution of Rodolfo's remains an entire month after it had been ceded to the anatomists of the *Sapienza*: "Il cappellano referse al Provveditore essere stato rimandato da notomisti il corpo o fussero le ossa e carne del corpo che si dette al 15 gennaio passato di Rodolfo Bernabeo per farne notomia e che era stato sepolto nel luogo solito." A similar case is reported in the Modenese chronicle of Jacopino de' Lancellotti for March 7, 1494: "Fu apichato uno el quale aveva asasinato zente e morte e robate e era ordinato de squartarelo non li fu el m.o [maestro] che lo squartasse, poi fu dato a li medeci li quali lo portorno in contrada de san Zohano vecchio incaxxa de queli dal Banbaxo [medico] e li fu smembrato e fato notomia" (quoted from P. Di Pietro, "Contributo alla storia degli studi anatomici in Modena," in *Atti e memorie della Deputazione di Storia Patria delle antiche Provincie Modenesi*, series 8, vol. 9 (1957), pp 81–87.

quartering. In the words of the recorder of the entry, it too assumed the aspect of a supplementary punishment that was being inflicted on the criminal after death—a punishment that, in this light, is heavy with significance: if quartering indicates a wish to taint the cadaver with the mark of infamy, the same can be achieved by dissection.

Through this textual fragment, in which the usual terseness with which the charitable acts of the Confraternity of San Giovanni Decollato are recorded is set aside, it seems to me that an image of the cadaver destined for dissection is evoked that emphatically contradicts the one that the confraternity itself and the other actors involved in the performance of the anatomy had wished to provide by way of regulatory texts and the ritual practices that preceded and followed the public anatomy lesson. This disreputable image of the dissected body was close to the view held by a wider audience, by those who, lacking familiarity with the true importance of anatomy, only perceived its violent and profaning aspects.

The dialectic, in fact the conflict, between scientific reason and cultural and religious constrictions was at the basis of the entire problem of dissecting practice and manifested itself most forcefully over such points as the choice of the cadaver, its display, its dismemberment, and delayed burial. The more strictly normative and institutional aspects of the public anatomy lesson temporarily defused its negative connotations, erecting lawful but circumscribed limits to the acquisition of scientific knowledge. However, anatomy was able to develop within them and to enhance its status as a discipline, while dissection found its own legitimization.

CHAPTER THREE

Tradition: An Archeology of Anatomical Knowledge and of Dissecting Practices

Any reader of Renaissance texts on anatomy will quickly notice their obsessive repetition of the names of physicians and scientists who had discussed human and animal morphology in earlier periods. It follows that the works of Aristotle, Hippocrates, and Galen especially were the obligatory sources and references for fifteenth- and sixteenth-century writers. Also mentioned are numerous Alexandrian, Byzantine, Arab and medieval Christian physicians, anatomists, and biologists who to varying degrees are part of the tradition and are integral or related to how anatomy has been conceived, practiced, recorded, and taught for much of the modern period.

Writers on anatomy did not restrict themselves to using and citing the works of their predecessors and of ancient authorities. In the prefaces and occasionally in the very body of the text they reconstructed a history of this science that recalls the fundamental stages, real and legendary, of its centuries-old development. This had a triple objective: first, to use the historical account to show the utility of anatomy for the evolution of medical science and, more generally, of natural philosophy, providing evidence from as far back as antiquity covering the possible applications of accumulated knowledge about the human and animal body; second, to explicate the position of the author with respect to that of the discipline (almost always suggested in a rather apologetic manner); and finally, to provide through this tradition itself, bolstered by ancient "auctoritas" and corroborated by a millenarian history, an epistemological and anthropological legitimacy for their activity as authors and especially as anatomists and dissectors.

The anatomical histories that they relate roughly resemble one another.

The "auctores" cited, the episodes recalled, the discussion pursued, and even the language used are frequently similar, and demonstrate how much they had all depended on the same small number of passages and episodes taken from the works of an even more limited array of authors (especially Galen, Celsus, Pliny, and Oribasius). Moreover, it seems obvious that from the end of the fifteenth to the beginning of the sixteenth century a diagrammatic summary of the most significant stages in the history of anatomy was tacitly drawn up, and whoever wrote on the subject based himself on it, plagiarizing contemporary writers, forgetting the original sources, and sometimes misrepresenting or even falsifying their contents.

In this chapter I do not intend to trace the intricate network of borrowing among the authors of these treatises but rather to reconstruct the salient characteristics of the anatomical tradition through an analysis of the essential themes and anecdotes that most interested Renaissance anatomists. What I envisage is a journey back to the ancient and medieval texts and other evidence that formed the basis of modern anatomy. My purpose is to reconstruct a sort of genealogy of the key concepts on which Renaissance anatomical knowledge was established and of the practices resulting from these epistemological and philosophical formulations. By reestablishing a chronology of the anatomical tradition, the reasons for the theoretical foundations of anatomy and each individual technique pertaining to it can be identified, along with the conditions that made this science possible and the devices adopted for its study and instruction. First among them is the one that was scientifically the most obvious and anthropologically the most risky: human dissection.

PHYSICIANS AND PHILOSOPHERS WORKING ON THE DISCOVERY OF THE BODY, OR THE USES OF ANATOMY

The first anatomical dissections of which there is any record were performed in Alexandria during the third century B.C. On this point there is agreement between contemporary scholarship and the stories told by Renaissance anatomists. The principal sources of this information are the writings of Galen and the *Proemium* to the *De Medicina* of Aulus Cornelius Celsus (42 B.C.–37 A.D.).[1] This last text is of particular interest inas-

1. There are countless printed editions of Celsus's work. The most frequently used version was prepared by F. Marx and appears in the *Corpus Medicorum Latinorum I*. For the *Proemium* (henceforth abbreviated as *pr.*) I shall cite from the critical edition by P. Mudry, *La préface du 'De Medicina' de Celse* (Rome, 1982). Doubts had been raised concerning Celsus's assertion that dissection and vivisection had been practiced in Alexandria, but the accuracy of his statement has been confirmed by L. Edelstein, "The History of Anatomy in Antiquity," in *Ancient Medicine*, ed. O. Temkin and C. L. Temkin (Baltimore, MD 1962),

much as the author provides us with a brief history of medicine up to his day, paying special attention to the various characteristics of medical practice in the opposing Empiricist and Dogmatist (or Rationalist) schools. The former, who were resolute adversaries of dissection, based their view of the healing powers of medicine on experience acquired by observing disease and the effects of attempted remedies. The second faction favored dissection of the human body, presupposing that practice and experience alone could not be enough to cure disease if a prior solid knowledge of the individual organisms and their nature was not present.[2] Celsus made himself the impartial spokesman of each of the schools and only at the end clarified his own position and his opinions on the epistemologies described, taking great care to distinguish them from the passages where he reported the views of others. Although it is not the principal contentious issue in the quarrel (not even in Celsus's *Proemium*), but only one of several matters of dispute, dissection presents certain knotty questions, essentially of a theoretical nature, that continually crop up in the history of the anatomized human body from the Hellenistic Age to the late Renaissance. Moreover, the passage from Celsus affords the possibility of going back even further, to Aristotle and to the Hippocratic writers.

If it is true that in the third century B.C. there were physicians in Alexandria who not only opened the bodies of the dead but occasionally even those of living condemned criminals, the first problem to confront, aside from the supporting contextual historical data, is what conditions of a medical and philosophical sort could have favored the utilization of techniques—dissection and vivisection—that have never previously been documented.

The physicians' objective was principally—to state the obvious—the curing of illness. For every good practitioner, beginning with the texts of the *Corpus Hippocraticum* of the fifth and fourth centuries B.C. until as late as the seventeenth century, healing was achieved by establishing or reestablishing a proper equilibrium among the humors which, when in a state of disharmony, produced pathological conditions in the human body. For an extremely long span of time covering the entire history of

---

pp. 274–301 (originally, "Die Geschichte der Sektion in der Antike," *Quellen und Studien zur Geschichte der Naturwissenschaften und der Medizin* 3, no. 2 [1932–33]), and, more recently, by H. von Staden, *Herophilus: The Art of Medicine in Early Alexandria* (Cambridge and London, 1989), especially pp. 45–48.

2. " . . . alii sibi experimentorum tantummodo notitiam necessariam esse contendunt, alii nisi corporum rerumque ratione comperta non satis potentem usum esse proponunt" (Celsus, *De medicina, pr.* 12).

medicine until the eighteenth century, the theory of humors, in its diverse variants, was the paradigmatic response to all pathological manifestations, whether they afflicted the external or internal organs of the body.[3] If curiosity about the structure of the human body was to have a place within the field of medical concerns, it had to fit into the paradigm of the theory of humors. In other words, if, as Empiricist physicians insisted, the theory of humors had entirely sufficed, both to explain the origins of disease in general and to provide the specific prognosis and diagnosis, anatomy as an auxiliary science would not have been necessary. The anatomical structure only begins to be relevant when it becomes necessary to locate the parts of the body in which the humors dwelled or were produced as well as the way by which they could be transmitted from one part to another both under normal as well as under pathological conditions. Herophilus himself, expert dissector and author of an anatomy treatise,[4] was also a believer in and defender of the theory of humors, which shows that even then (as later in the Renaissance) there was no intrinsic contradiction between a belief in the humoral paradigm and anatomical research.[5] Moreover, Herophilus was not the only physician of antiquity operating in the context of the theory of humors who also had anatomical interests. A large part of medical writing, at least beginning with Alcmaeon of Croton (mid-fifth century B.C.), through the writers of the *Corpus Hippocraticum,* the Alexandrian rationalists, and Galen, was based on knowledge, however rough, of human anatomy. Even more compelling is the fact that though the greater number of works dealing

3. The four fundamental humors, as we know, were blood, phlegm, yellow bile, and black bile. They were equivalent roughly to the four elements: respectively air, water, fire, and earth. Although there was a long-standing agreement between physicians and philosophers on the principle of the correspondence between elements and humors, this was not the case when it came to the number and variety of humors. On this question, see, for a general panorama, H. E. Sigerist, *A History of Medicine* (Oxford, 1951–61), 2:317 ff.; the introduction by W. H. S. Jones to *Hippocrates,* 4 vols. (London and Cambridge, 1923–31), 1:XLVI–LV; and E. Schoener, "Das Viererschema in der antiken Humoralpathologie," *Sudhoff Archiv* 48, fasc. IV (1964).

4. Only three fragments remain of this treatise, which is now lost: the first from bk. II (in Galen, *De anat.* 6.8, K II 570–72 and, taken from it, Oribasius, *Collectionum medicarum reliquae,* 24. 25. 1–6, CGM, VI.2.1); the second from bk. III (in Galen, *In Hippocratis de Epidemiis libros,* 2.4.1 *Commentarius* 4, CGM, V.10.1); the third from bk. IV (in Galen, *De semine* 2.1, K IV 586–88). On Herophilus's anatomical treatise, see H. von Staden, *Herophilus,* pp. 153–81 and pertinent texts. Also on Herophilus and Alexandrian medicine, see P. M. Fraser, *Ptolemaic Alexandria,* 3 vols. (Oxford, 1986), 1:338–76 (1st ed. Oxford, 1972).

5. Celsus, *De medicina, pr.* 14–15: "Neque esse dubium quin alia curatione opus sit si ex quattuor principiis vel superans aliquod vel deficiens adversam valetudinem creat, ut quidam ex sapientiae professoribus dixerunt; alia, si in umidis omne vitium est, ut Herophilo visum est."

exclusively with anatomy written between the fifth century B.C. and the first century A.D. have been lost, some fragments have survived.[6]

All Hippocratic medicine and all the later scientists who adhered to it began with the proposition that diseases could not be cured without an appropriate prior acquaintance with the structure of the human body: the *De locis in homine* clearly states that "The nature of the body has to be the foundation of medical training." On the other hand, the *Corpus Hippocraticum* does not contain a single text specifically dedicated to the general description of human anatomy, with the exception of a Hellenistic fragment.[7] Nevertheless, in the *Corpus,* knowledge of the essential aspects of anatomy are implicit and operative in all the therapeutic and surgical texts.[8] The latter, especially, demonstrate an acquaintance with superficial anatomy and osteology: surgery, in fact, cannot do without topographic knowledge.[9]

6. For an exhaustive discussion of anatomy in antiquity, see L. Edelstein, "The History." Cf. G. E. R. Lloyd, *Magic, Reason and Experience: Studies in the Origin and Development of Greek Science* (Cambridge, 1979), pp. 146 ff. and, in particular on Alcmaeon of Croton, G. E. R. Lloyd, "Alcmaeon and the Early History of Dissection," *Sudhoff Archiv* 59 (1975): 113–47 (reprinted in Lloyd's *Methods and Problems in Greek Science* [Cambridge, 1991], pp. 164–93), in which the author demonstrates that Alcmaeon never practiced actual dissections on the human body, as others have maintained. The first anatomical treatise, according to what Galen states in his *De anatomicis administrationibus,* was written by Diocles, a contemporary of Theophrastus (fourth century B.C.) (*De anat.,* K II 282). On Diocles, see W. Jaeger, *Diokles von Karystos: Die griechische Medizin und die Schule des Aristoteles* (Berlin, 1938). Praxagoras of Cos, the younger, who was the conduit for Hippocratic medicine in Alexandria and the teacher of Herophilus, also appears to have been the author of anatomical texts. On Praxagoras, see the 1954 contribution by K. Bardong, "Praxagoras," in *RE,* XXII, 2, coll. 1735–43, and the article by Baumann in *Janus* (1937): 167–85. F. Steckerl (*The Fragments of Praxagoras of Cos and his School* [Leiden, 1958]) is much less precise on the biographical data. On Praxagoras's anatomy, see especially L. Garcia Ballester, "El saber anatómico de Praxagoras de Cos," *Bollettino de la Sociedad española de Historia Medica* (1966): 43–49.

7. E. Littré, VIII, pp. 538–40. The dating is the work of G. Harig and J. Kollesch, "Galen und Hippokrates," in *La collection hippocratique et son rôle dans l'histoire de la médecine,* ed. L. Bourgey and J. Jouanna (Leiden, 1975), pp. 257–74. The *De corde* (E. Littré, IX: 76–93) is another preeminently anatomical text, dating from the third century. It appears to be actually posterior to the anatomical writings of Erasistratus. See C. R. S. Harris, *The Heart and the Vascular System in Ancient Greek Medicine from Alcmaeon to Galen* (Oxford, 1973).

8. Not much has been written about the *Corpus Hippocraticum.* In addition to the article by Edelstein (cited at note 1), see H. Kuhlewein, *Die chirurgischen Schriften des Hippokrates* (Ilfeld, 1897–98), in particular pp. 8, 12–15; the somewhat superficial contribution by L. Stroppiana, "L'anatomia nel Corpus Hippocraticum," *Rivista di Storia della Medicina* 7 (1963): 9–17; and, more recently, chapters 11–16 in V. Di Benedetto, *Il medico e la malattia: La scienza di Ippocrate* (Turin, 1986), in which the role of anatomy is probably exaggerated.

9. For example, the treatises on head wounds, fractures, articulation, instruments of reduction, and the physician's workshop, all of a surgical bent.

Familiarity with anatomy, however, was not the exclusive preserve of physicians. We have noted that they functioned within the paradigm based on the theory of humors and used anatomy only as secondary, auxiliary knowledge or in surgical practice. On the other hand, the question of how the human body was made pertained as much to philosophy as to natural science, according to the ancient and the Renaissance texts. This is an extremely important point not only for the development of this discipline in antiquity, but also for fifteenth- and sixteenth-century authors, including Vesalius and his followers. Renaissance writers on anatomy never failed to remind their readers and students that the discipline had an allegiance to both camps, the medical and philosophical. Alessandro Benedetti, for example, in the dedicatory letter of his *Anatomice* to Emperor Maximilian, affirmed that anatomy was properly a branch of philosophy used by the medical art,[10] and Berengario, after declaring that anatomy was necessary for theoretical medicine (in searching for the causes of illness) and for practical medicine (in their cure), wrote: "The utility and necessity of anatomy not only is required knowledge for the physician but also for the philosopher probing the secrets of nature"; the latter, through anatomy, could admire the power of the creator.[11] Statements of this sort can also be found in the writings of Winther, Massa, and Coïter,[12] and even in the *Fabrica* of Vesalius; he defined this discipline "as belonging to natural philosophy" and stressed that the study of the detailed structure of the human body should be undertaken "for itself," independently of its relevance to the field of medicine.[13]

This twofold application for the study of anatomy has very ancient roots and echoes what Galen wrote in his *De anatomicis administrationibus:* "Some [aspects of anatomical knowledge] are even of greater service

---

10. A. Benedetti, *Anatomice sive historia corporis humani. Ejusdem collectiones medicinales seu aphorismi* (Venice: Bernardino Guerraldo Vercellensis, 1502).

11. I. Berengario da Carpi, *Commentaria*, fols. vr–v.

12. Johann Winther maintained, taking his cue from Galen, that knowledge of anatomy was useful for all men: " . . . deinde quod Anatomae cognitio non tantum medicis ac philosophis utilis et necessaria, sed etiam omni hominum generi et honesta et pulchra hodie neglecta prorsus iaceat" (*Institutionum anatomicarum*, p. 3). Niccolò Massa, in his *Liber introductorius,* fol. 3v, refers to Galen without naming him ("utilis et necessaria philosophis ac pariter medicis sit") and goes on to state that the interest of anatomy lies in discovering the secrets of the most perfect creature in nature ("homo a natura ultimus intentus ultimam naturae perfectionem ostendat"). Folker Coïter (*Externarum et internarum principalium humani corporis partium tabulae,* 1573, fol. AA2r) distinguishes the anatomy of the philosophers, the first beneficiaries of this discipline, "utilis . . . per se," from the "necessary" one of the physicians. He adds: "Anatome scientiae verae Dux est aditumque ad Dei O.M. omnipotentiam ac iustitiam, quibus in construendis et formandis animantium corporibus usus est, proebet:" anatomical investigation provides access "in Dei cognitionem."

13. A. Vesalius, *De humani corporis fabrica*, fols. 3r, 4v.

to philosophers than to physicians," emphasizing "that one thing is the utility of anatomical theory for the naturalist who loves science for itself, and another thing for anyone who does not love it for itself but to show that nature has done nothing in vain," obviously referring to philosophers.[14] It has often been pointed out how ephemeral the boundaries were between philosophy and science in antiquity on the theoretical, epistemological, and cultural level, as well as in the ensuing reflections of this on later practices. In an important article from the 1930s Ludwig Edelstein emphasized the anatomical activity of philosophers, revealing both the motivation for this activity as well as the ways in which it took place. Anatomy for philosophers followed a course parallel, chronologically, to the rise of a systematic interest in the human body on the part of physicians.[15] Without going back as far as Anaximander and to much of pre-Socratic philosophy, which was dedicated to the acquisition of knowledge about the composition of the human and animal body, one need only consider Plato's *Timaeus,* and especially the writings of Aristotle (*The Parts of Animals, The Reproduction of Animals, The History of Animals*). It is common knowledge that these texts continued to be quoted and used in anatomy treatises from Herophilus until at least Vesalius.

In Plato's *Timaeus* physiology is considered entirely as a philosophical matter (quite distinct from any medico-scientific ones). In the *Timaeus* Plato set out to prove divine power by contemplating the mechanics of creation. Aristotle's writings differed not so much in substance as in method: Aristotle was constantly preoccupied with eminently philosophical questions, or rather, ones of natural philosophy, and he enters into empirical research to explain and interpret the way the world works. Inevitably included in these concerns was the morphology and the composition of bodies and their parts. The method he applied, as can be seen from the zoological texts, was that of direct observation through a repeated comparative dissection of animals.

At this point the major distinction between the anatomy of the physician and the anatomy of the philosopher is manifested by the use of this knowledge in their specific fields. The anatomical knowledge of the physician was directed toward the healing of diseases afflicting internal organs (when possible) and exterior parts of the body and to the practice of surgery; it was reduced to the essentials (a superficial topographical proficiency) by an extremely powerful etiological, therapeutic, and to a certain degree, self-sufficient paradigm. The anatomical learning of the philoso-

---

14. Galen, *De anat.,* K II 286 ff.
15. L. Edelstein, "The History," especially pp. 260–73.

pher on the other hand was directed to the verification and to the mastery of the principles that inform man and nature.[16]

Galen, commenting on the diverse possible uses of anatomy, recalled which aspects of the discipline were essential for the physician: "all the parts of the arms and legs, and all the parts, *external* rather than *internal,* of the shoulders, back and chest, of the ribs, abdomen, neck and head."[17] The rest he defined as the "excess," the "superfluous" that one needs to be concerned about "because of the sophists, who are not satisfied with having known the art of nature in useful things, but every time bring up the problem of the purpose for which a certain thing has been made, because it was made like this or this big." This "excess" of knowledge, the anatomy of philosophers, served them "either for simple theoretical interest or to communicate that the art of nature succeeded in every part."[18] It was precisely the aspect of research "in the individual parts of the body," freed from an etiological, diagnostic, or therapeutic purpose, that gave to the anatomy of the philosopher a wider field of inquiry; although it was always conceived as a tool and not justified as a branch of knowledge to be pursued for its own sake, anatomy nevertheless found in natural philosophy a means of developing beyond the boundaries imposed by the extremely limited and circumscribed queries of medicine. It seems as if the perspective drawn from the philosophers' anatomy was much more open to empirical research and to factual verification than that which was drawn from the physicians'. The philosophers were far more likely to have questioned the unknown and the invisible (for example, the internal organs of the body and their functions) than were the physicians, who were restrained by their own epistemological paradigm.

This twofold evolution in anatomical research found its preliminary synthesis in Galen's *De anatomicis administrationibus,* even though the author distinguished the anatomy of parts useful to the physician from those useful in themselves or as aids in the theoretical speculation of the philosophers. Traces of this binary development remained in the arguments put forward by Renaissance authors, although they ceased to maintain any sort of distinction between the parts of the anatomy useful to the philosopher and those useful to the physician. They maintained that it was equally necessary for both disciplines to know the human body from

16. Ibid, p. 265.
17. Galen, *De anat.,* K II 284. A few pages before he reiterates that only this part of anatomy, and not the muscles, nerves, arteries, and veins of the heart or of the viscera, is useful to physicians (p. 287).
18. Ibid, K II 284–85 and 287. On the matter, see the summary but illuminating discussion in M. Vegetti, *Il coltello e lo stilo* (Milan, 1987), pp. 39–42.

every angle. Only the use to which this knowledge was put differed; philosophers, along with physicians, were ranked among the inventors of the anatomical art.

Unveiling: Dissecting Animals, Dissecting Humans

A large part of the controversy in which the supporters of human dissection in the sixteenth century, especially Vesalius, were engaged was directed against those who had described the parts of the body in long and prolix treatises and had used techniques other than the opening and observation of cadavers. This often resulted in grievous errors. These techniques, which continued to be employed at the same time as dissection (even during the sixteenth century), either because of the shortage of cadavers or because of the restrictions imposed by university regulations, provided the first attempts at anatomical investigation. It seems appropriate to trace here the evolution of anatomy from a practical point of view through the available Renaissance texts and treatises, with a double aim: first to look at the process that led to human dissection as the principal device for anatomical investigation, second to identify the theoretical premises on which these techniques were based and the type of question that it was believed they could answer.

In antiquity the acquisition of knowledge of the external structure of the human body did not pose a problem for physicians and natural philosophers.[19] However, the study of the interior parts presented a different challenge, one which, as we have seen, Galen deemed "excess" knowledge as far as its relevance to medicine was concerned. The inner body, hidden from view by the epidermal mass, constituted the limits of knowledge for ancient empirical research.

The opening up and observation of the inner human body seems to us the most obvious of available techniques for anatomy. But this was not

---

19. L. Edelstein ("The History," p. 261) cites on the subject two especially significant references broadly separated in time. The first comes from Aristotle's *Historia animalium:* "Above all one must observe the parts of man. Just as everyone judges coins on the basis of those we know best, so it is for all things. Man is necessarily among all living things the one with which we are most familiar" (Aristotle, *HA,* 491 at 19 ff.); the second comes from Gregory of Nyssa: "Sed enim unusquisque nostrum, natura propria magistra et duce, structuram corporis accurate ex eo quod et videt et vivit et sentit, cognoscere poterit" (Gregory of Nyssa, *De opificio hominis, PG,* XLIV, coll. 239–40). Both impart how much knowledge of the human body, at least in its visible parts, was a knowledge that could easily cross over specific professional boundaries and become appropriated by a more widely diffused culture. We should also bear in mind the importance of superficial anatomy in classical Greek sculpture and Homeric poetry, in which the minute descriptions of the wounded parts of heroes' bodies conveyed not only the author's detailed observations but also the interest of his public for this type of subject matter.

the case for scientists until the time when the school of Alexandria was founded. It appears that before the end of the third century B.C. such an operation was not only never performed but was not even conceived as possible. Texts never refer to the dissection of the human body, not even regretfully as to a desirable but forbidden procedure. Thus physicians and philosophers worked out other more indirect and (later anatomists would say) fallacious techniques through which to pursue their investigative purposes. These methods were basically two: according to the first, mentioned in certain writings of the *Corpus Hippocraticum,* it was possible that certain characteristics not only of pathology but even of internal morphologies could be deduced from an external observation of the human body and its *palpatio;*[20] the second, considerably more common and valued, consisted of the dissection and vivisection of animals that more or less resembled humans. Each of these techniques made a supplementary assumption: the first, that there was an adequate semiological procedure and that the inferring of the interior of the body from the exterior was theoretically valid; the second, that deducing human structure from that of the animal was valid, and that a theory could be created that would allow the use of conceptual tools such as analogy and homology.

Though inferring the internal parts from the exterior of the body was only momentarily successful, since as a method it showed itself to be untrustworthy and imprecise, the dissection and vivisection of animals continued to be practiced vigorously during the entire history of anatomy. The first definite evidence of animal dissection goes back to a period between the end of the fifth and the early fourth century B.C., to the years during which "On sacred disease" of the *Corpus Hippocraticum* was written.[21]

20. At least initially it had as its proper purpose that of following internal pathological processes through the observation of changes perceivable from without. But, according to Hippocrates in his *On Ancient Medicine,* certain anatomical characteristics of even some internal organs have to be deduced from the outside, although this presents notable difficulties (Hippocrates, *L'antica medicina,* chapter xxii, in his *Opere,* trans. M. Vegetti [Turin, 1964], p. 187). An invisible interior is juxtaposed to a visible exterior. The consistency, form, movements, and functions of some of these organs are summarily described in chapter 21 of this text and have evidently been ascertained through touching. For a fuller discussion, see G. E. R. Lloyd, *Magic,* p. 148. Another method of observing the vascular system from outside was the one frequently suggested by Aristotle, namely that of examining extremely emaciated living humans, in whom blood vessels were as visible as the veins of leaves (Aristotle, *HA,* 511 b 13 ff., 513 a 12 f. and 515b 1, but also *PA,* 668 a 21 f.).

21. E. Littré, VI: 350–97. Chapter 3 is dedicated to a curious description of the blood vessels that extend from the brain to the rest of the body, revealing, in spite of erroneous contents, considerable interest concerning the structure of the circulatory system. The description begins with the presupposition that a close morphological relationship exists between the form and structure of human and animal brains. On this basis, the author, in chapters 4 and 5, mentions the dissection of goats' brains for the purpose of investigating

In all those cases in which the Hippocratic texts mention the dissection of living or dead animals the analogy between man and animal is understood as a given: no theory, however, offered a clear legitimation of it.

It was undoubtedly Aristotle in his zoological writings, particularly his *Historia animalium, De generatione animalium,* and *De partibus animalium,* who established a general method for the observation of animate bodies through dissection and vivisection.[22] Moreover, he was recognized by Renaissance anatomists as one of the inventors of dissection and as being among the founders of anatomical science, together with Alcmaeon, Plato, Hippocrates, and Galen.[23] Very roughly, Aristotle's purpose could be said to be the classifying of all animals on the basis of an in-depth knowledge of their anatomical characteristics: "In a certain sense, it is through the configuration of the parts and of the entire body, when it presents similarities, that species can be defined."[24] The investigation and the ensuing classification are both principally intended to illustrate the four types of causes in the animal world and, in particular, the formal one and the final one, in regard to the whole body and its separate components.[25] This called for empirical research based not only on reasoning but also on direct observation.[26] The writings of Aristotle demonstrate a constant and methodical application of dissection as an essential tool for the disclosure of the internal parts of the animal body. The adoption of this method, however, raised problems of an epistemological and cultural

the causes of madness (*malattia sacra*) in man. In the *Corpus Hippocraticum* there is a reference to the dissection of a goat even in the *De corde;* anatomical observations on animals are recorded in at least two places. An experiment is described that involves slitting the throat of a hog while it is drinking and the removal and dissection of the heart of an unidentified animal (E. Littré, IX: 80, 88). For a detailed account of animal dissection in antiquity and a discussion of the historiography on the subject, see G. E. R. Lloyd, *Alcmaeon,* and G. E. R. Lloyd, *Magic,* pp. 146–69.

22. To these two treatises I should add *De motu animalium* and *De incessu animalium.* I have used the Italian edition: Aristotle, *Opere Biologiche,* a cura di M. Vegetti e D. Lanza (Turin, 1972).

23. See, for example, chapter III (*De artis inventoribus*) in the introduction to Coïter's *Externarum,* fols. AA2r. Berengario da Carpi includes him among the *auctoritas* of anatomy together with Galen and Avicenna (I. Berengario da Carpi, *Commentaria,* fol. IIIIr). However, it is enough to examine all the anatomical treatises written and published between the end of the fourteenth century and the sixteenth century to realize how often, for better or worse, Aristotle was cited.

24. Aristotle, *PA,* 644 b 7 ff.

25. Aristotle wrote at the conclusion of his treatise on the parts of animals: "We have thus dealt with parts, and we have set forth for what reason each one is to be found in animals" (ibid, 697 b 27 f.).

26. "Certain subjects need to be clarified by reasoning, for others, however, it is preferable to have recourse to direct observation" (ibid, 680 a 2–4).

nature. In his *De partibus animalium* Aristotle had, in fact, proposed a problem that would later be taken up by the detractors of the Alexandrian Dogmatic school in denying the utility of human dissection: the body of a dead animal was not equivalent to a living one; it conserved only its exterior form, not its substance, and not the vital spirit.[27] So it was necessary to open up and observe not only dead animals but also living ones. Furthermore dissection, whether of the living or the dead, could be considered "base" because of a Pythagorean taboo, which perhaps also motivated the hesitation of Hippocratic physicians to assume dissection as a research tool.[28] Aristotle overcame both obstacles and demonstrated a willingness to kill for no other purpose than the acquisition of knowledge,[29] adducing an aesthetic justification in the area of a naturalist's research: "We must not feel an infantile repulsion toward the study of the most humble living beings: there is something marvelous in all natural things."[30]

Aristotle's works witness the fact that the dissection of animals was practiced before him and his collaborators, but everything we know suggests that these were superficial operations, performed rather sporadically, dictated by specific circumstances and not by a methodology solidly rooted in theory.[31] This has been confirmed even by the most recent scholarship, which contradicts what the anatomists of the Renaissance and even Galen could not help believing: namely, that before Aristotle, the divine Hippocrates had performed dissections, just as Plato had (in order to write the *Timaeus*) and Alcmaeon (for his research on the eyes). For Aristotle dissection was not an instrument used occasionally as a response to incidental questions, but rather a methodology firmly founded on a theory capable of disclosing the secrets of nature and the truth.[32] If further proof is necessary it is provided by the fact that Aristotle on various occasions mentions a treatise of his, *De dissectione*, now lost, which was specifically dedicated to the subject and which was replete with illustra-

27. Ibid, 640 b 34–35. And he adds: "no part of a cadaver, for example an eye, a hand, is any longer truly such" (ibid, 641 a 4–5).

28. Ibid, 645 a 27 f.: "If then someone should consider unworthy the observation of other animals, then he should judge his own in the same way." On the Pythagorean taboo, see M. Vegetti, *Il coltello*, pp. 19–21, 38.

29. Ibid, p. 30.

30. Aristotle, *PA*, 645 a 15 ff.

31. "Those who have conducted observations on dead and dissected animals have not seen the origins of the principal veins, while instead those who have conducted them on living emaciated humans have succeeded in identifying the origins of veins on the basis of what they could see externally" (Aristotle, *HA*, 511 b 13 ff). Cf. G. E. R. Lloyd, *Alcmaeon*.

32. On the relationship of dissection to truth in Aristotle's zoological works, see the excellent account in M. Vegetti, *Il coltello*, pp. 57–61.

tions and anatomical diagrams.[33] The evidence would indicate that it consisted of an instructional manual for the *Lyceum*.

Man is naturally included among the animals. Aristotle asserts this constantly and at various points makes man the purpose of the comparison: "Among the animals known to us, man is either the only one to participate in the divine, or the one who participates in it in the greatest measure. For this reason, thus, and also because the form of his external parts is the best known, we shall have to speak of man first of all."[34] However, man is considered here merely as one of the many animate creatures of nature and is not the object of greater research attention than the others. But he does occupy a privileged position from the methodological point of view, since he is certainly the most complex of the animals, and, at least in his external parts, the most familiar. It was not possible though to obtain a direct knowledge of the internal parts of the human body, as it was for other animals, through dissection. If the dissection of animals was considered shameful by the many, "it is in fact not without a great disgust that we observe what the human species is composed of: blood, flesh, bones, veins and similar parts."[35] The very impossibility of conceiving the opening up of the human body did not, however, limit the philosopher's investigations: man's interior parts, still so little known in the fourth century, did not belong to the world of the unknowable as much as did those of other animals: "They must be studied with reference to the parts of the other animals whose nature resembles man's."[36] This was especially true in regard to the monkey, defined by Aristotle as a caricature of man,[37] but it also applied to a number of other complex animals.

The first explicit exposition of a methodology destined for long-term success begins with Aristotle. Moving from Galen to the Renaissance it crossed the centuries to arrive at the experiments still being conducted today, which are based on the analogy between human and animal morphology and physiology. This analogy was the essential conceptual device

33. See, as examples, Aristotle's references in *HA*, 497 a 32, 525 a 8 ff., 566 a 14 ff.; *GA*, 740 a 23, 746 a 14 ff.; *PA* 650 a 31 ff., 668 b 29 ff., 674 b 16 ff., 680 a 1 ff., 684 b 4 ff.

34. Aristotle, *PA* 656 a 7 ff. As G. E. R. Lloyd has shown, for Aristotle, in addition to being part of the investigation, man is also considered the starting point of reference in his compilation (*Science, Folklore and Ideology: Studies in the Life Sciences in Ancient Greece* (Cambridge, 1983), especially chapter 3 of pt. 1, "Man as a Model," (pp. 26–43).

35. Aristotle, *PA*, 645 a 24 ff.

36. Aristotle, *HA*, 494 b 21.

37. *Topica*, 117 b 17 ff.: "In fact, if the resemblance is not of a caricatural sort, such as the one that unites the ape to man. . . ." On the ape in ancient medicine and philosophy, see M. Vegetti, "L'animale ridicolo," in M. Vegetti, *Tra Edipo e Euclide: Forme del sapere antico* (Milan, 1983), pp. 59–70, and W. C. McDermott, *The Ape in Antiquity* (Baltimore, 1938).

that enabled Aristotle to gather data on the internal parts of the human body. Of course it was intuitively evident, but Aristotle made it his epistemological focus, and it allowed him to fashion an empirical, systematic research methodology.[38] "By analogy I mean that certain animals have a lung, others do not have a lung but some other organ that fills the role of the lungs in animals who have it; another example, some have blood, others something analogous that possesses the same properties as blood in sanguine animals."[39] Basically Aristotle's analogy was an analogy of functions rather than a homology of structures. On this basis, by studying the organs of certain animal cadavers ("more similar to man's") and relating them to the corresponding human organs, the existence of which was speculatively or semiologically verifiable, it was possible to infer from the first the functioning of the others. If, on the one hand, this procedure suggested a functional uniformity in nature, on the other it posed the problem of the differentiation of genera and species for the purpose of a classification based on structures. Aristotle seemed to be especially interested in defining and illustrating those elements that set man apart from the rest of the animal world:[40] he did this not so much by way of analyzing the function of the internal parts but largely through the confirmation of superficial differences.

A comprehensive and exhaustive reconstruction of the structure of the human body cannot be derived from such an analogical process, nor even less from an approximate description of those organs capable of controlling what Aristotle considered the essential functions of the body: reproduction and survival.[41] Aristotle's zoological works, however, made two fundamental and lasting contributions to anatomical methodology. First, they proposed the systematic practice of dissection as a tool for learning about the interior parts of animate bodies and their functions: the evidence obtained from observation is placed first among the principles of naturalistic knowledge. Second, they provided a theoretical legitimation of a research paradigm that was based on the analogy between animal and human physiology.

38. On the analogy, see G. E. R. Lloyd, *Polarity and Analogy: Two Types of Argumentation in Early Greek Thought* (Cambridge, 1966).

39. Aristotle, *PA*, 645 b 6 ff.

40. On this issue, see G. E. R. Lloyd, *Science*, p. 35.

41. Guided by the suggestions of Aristotle, some of his students broadened the field of investigation and, moving from the study of vital organs, dedicated themselves to the observation and arrangement of other parts of the body, as is demonstrated, for example, by the treatise on skeletons of Clearchus of Soli, which has not survived even in fragmentary form. It too was probably based on direct observation through dissection and presumably also had a taxonomic character.

Neither Aristotle's empirical research, founded on the principle of a functional analogy of the internal parts of the body, nor medical practice inspired by Hippocrates between the fifth and third century B.C., which based its diagnostic and therapeutic method on the theory of humors and on an external examination of symptoms and pathological processes, were to lead to the formulation of a theoretically coherent question that would require the opening of the human body. It is nevertheless clear that the practice of animal dissection and a broadening of the field of medical inquiry would shortly lead to working on the human body, since only that would satisfy the new demands. The dissection of animals had revealed the possibilities offered by direct observation. The investigations of physicians would increasingly become oriented toward detailed research, especially of the sensory organs, the nervous system, and the circulation of the blood. The opening of the human body as a research tool was the next step.

This occurred in Alexandria in the third century B.C., according to Celsus, who reported it some centuries later.[42] There a medical school came into being that was recalled in subsequent centuries as a sort of Eldorado of anatomical research. Vesalius, for example, looked forward to an age when universities would return to that happy time when a knowledge of real anatomy was cultivated, as it had been in Ptolemaic Alexandria. That period was contrasted to the rest of the history of the study of human anatomy, which seemed to be derived from a slavish handing down of frequently false facts acquired from an observation of animals and especially from earlier, supposedly authoritative texts.[43] This celebrated school, free of prejudices, had Herophilus as its greatest representative and served as a link between the most significant medical traditions of antiquity, the Hippocratic and the Galenic.

Few of the names of the physicians who flocked to Alexandria have come down to us, but most seemed to have been from Cos, the island where the Hippocratic school originated and developed. Even Herophilus of Chalcedon had studied at Cos under Praxagoras, one of the last masters of the Aesculapian tradition. Not only did he follow its theories both in his practice and teachings, but he also promoted the critical study (especially in the form of glosses and commentaries) of those texts that we now call the *Corpus Hippocraticum*.[44] Unfortunately, none of these com-

---

42. On Alexandrian culture during this period and on the role of its ruling family, see the large work by P. M. Fraser, *Ptolemaic Alexandria*, 3 vols. (Oxford, 1986; 1st ed. Oxford, 1972), with a full bibliography on these subjects.

43. A. Vesalius, *De humani corporis fabrica*, fol. 3v.

44. They had reached Alexandria thanks to certain physicians from Cos who had brought portions of their libraries with them. See P. M. Fraser, *Ptolemaic Alexandria*, pp. 346 and 364 ff.

mentaries, nor any of Herophilus's original works or even those of Erasistratus, have survived. By the Middle Ages and the Renaissance only sporadic, indirect pieces of evidence and a few brief fragments that can be
found in Galen's works and in this or that Byzantine commentator have
survived. Thus it is difficult to give a considered historical evaluation of
the anatomical activity of Herophilus and Erasistratus, to discuss the contents of their treatises, or to corroborate the opinion of Celsus through
an analysis of their objectives and methods and of the expectations placed
on them.[45] At any rate, it seems clear from the subsequent abundant
evidence that anatomy owed much to the Hellenistic physicians and especially to Herophilus, whose fame extended without a break to the
sixteenth century, when Gabriele Falloppio called him the Vesalius of
antiquity, both for his ruthless practice of dissection and for his powers of
observation. Herophilus certainly wrote an anatomy treatise; numerous
discoveries and descriptions that later writers passed down to us are attributed to him.[46] These are the sole surviving traces of his work, but even
they suggest a conception of medicine in which empirical research and
knowledge of human anatomy, both its internal and its external parts,
assumed a much larger role than it had previously. The only way in which
he could answer the questions that were the objects of his research was
by direct observation of the human body. We do not have enough information to judge what the epistemological assumptions were that impelled
him to adopt the technique of dissection. We are even less able to judge
in what way they differed from the assumptions of those who wrote on
human anatomy before him (Diocles of Carystus and Praxagoras of Cos).
We unfortunately lack the sources that might have helped us to establish
the terms by which dissection was justified in Alexandrian anatomy.

Even greater difficulties seem to arise in the case of Erasistratus of
Ceos.[47] There is no evidence, in fact, that he ever wrote a treatise of de-

45. Herophilus's fragments have been published by H. von Staden, *Herophilus;* for those
of Erasistratus, see I. Garofalo, *Erasistrati Fragmenta* (Pisa, 1988).
46. The anatomy of the brain and the identification of the cerebral ventricles, the discovery of the nerves, the description of the membranes of the eye and of the human liver, the
identification and measurement of the duodenum, the first generic distinction of arteries
from veins, a careful differentiation between the various parts of the spermatic ducts, the
discovery of ovaries and at least of a part of the Fallopian tubes, the discovery and labeling
of certain vascular structures, such as the *torcular Herophili.* Cf. H. von Staden, *Herophilus,*
pp. 155 ff. and apposite texts 25–129.
47. It appears that there is no obvious evidence for an Egyptian sojourn on the part of
Erasistratus. It has been hypothesized, instead, that he was active in Antioch, during the
reign of the Seleucids, a dynasty competing with the Ptolemys both in the political sphere
as well as for cultural primacy in the Mediterranean world. See P. M. Fraser, "The Career
of Erasistratus of Ceos," *Rendiconti dell'Istituto Lombardo.* Classe di Lettere e Scienze
Morali e Storiche, 103 (1969): 518–37.

scriptive anatomy, and his account of the internal parts of the human body were contained in physiological and therapeutic works that are now lost. The few surviving fragments and the information that can be extracted from subsequent writers seem to indicate that Erasistratus was successful in the first major attempt to construct a theory of disease as well as a clinic, both based on anatomy, whose failures were quickly attacked by the Empiricist school. Information on Erasistratus's activity as an anatomist can be extracted generally from fragments embedded in the works of Galen: they provide certain evidence of a recourse to autoptic practices to determine the causes of death, and the description of the brain and of the cranial nerves in man are witness to the fact that Erasistratus certainly conducted research to construct a descriptive anatomy.[48] But these are just crumbs of a whole research program, the basic premise of which remains hidden. What is certain, however, is that both Erasistratus and Herophilus were physicians educated in the Aesculapian school and that whatever they produced as anatomists must unquestionably be considered as a part of this tradition that proclaimed itself to be Hippocratic, as was confirmed first by Celsus and later by Galen.[49]

The extreme scarcity of original writings thus constitutes an insurmountable obstacle to the reconstruction of the motives that led Hellenistic physicians to take up the dissection and vivisection of the human body. We are obliged to consult later authorities, especially Celsus: it is precisely through the filter of the *De Medicina* and then through the writings of Galen, Rufus, and Soranus that the reputation of Herophilus and Erasistratus, as the first practitioners of dissection in the history of medicine, was handed down to the sixteenth century. They are portrayed there as fearless heroes of science or as despicable butchers.

In Celsus's few surviving pages, allusions to arguments both for and against the opening of the human body advance an array of opinions that, now explicitly, now tacitly, were from the very beginning at the center of the debate over the centuries. In the part of the *Proemium* of the *De medicina* dedicated to the exposition of Dogmatist doctrine, Celsus stated the objectives of Dogmatist (or rational) medicine: first of all, to know the hidden causes that determine the pathological states, the evident causes, the natural functions (respiration and digestion, for example) and the

---

48. On anatomy in the works of Erasistratus, see I. Garofalo, *Erasistrati Fragmenta*, pp. 22–29. For autopsies, in particular, see fragments 251, 280; for the anatomy of the brain, fragment 289.

49. Herophilus and Erasistratus were in fact enrolled in the ranks of rational or Dogmatist physicians: see Galen, *Introductio seu medicus*: "Hippocrates of Cos was founder and head of the sect of rationalists. He was followed by Diocles of Carystus, Herophilus, Erasistratus. . . ." (my translation). Cf. Celsus, *De medicina, pr. 8.*

purposes of the individual internal organs.[50] The study of natural functions especially, was, in Dogmatist medicine, an indispensable prerequisite for combating and healing the diseases that affected the body's mechanisms. Similarly, it was considered impossible to cure the disorders of internal organs without an adequate knowledge of them,[51] even in cases where these afflictions had to be treated with external remedies.[52] In view of these exigencies, to Dogmatist physicians, heirs of Herophilus and Erasistratus, there seemed to be no other road to take except that of the dissection of the human body. Neither the inductive process of the Hippocratics nor the analogical one of the Aristotelians sufficed any longer. Research on the human body had to be precise and meticulous so as to unveil "those parts which nature had previously concealed." Of the individual organs, it was necessary to know "their position, color, shape, size, arrangement, hardness, softness, smoothness, connection, and the projections and depressions of each, and whether anything is inserted into another thing or whether anything receives a part of another into itself."[53] Any chance of using alternative techniques of discovery was quashed by the rise of these new cognitive objectives. "Therefore," Celsus continued, "it is necessary to dissect the bodies of the dead and to examine their viscera and intestines."[54]

But Dogmatist physicians did not stop here. In the opinion of Celsus, "Herophilus and Erasistratus, they say, did this in the best way by far when they cut open men who were alive, criminals out of prison, consigned to them by kings."[55] To open up living bodies meant being able to observe internal organs while they were still functioning; a beating heart, pulsating veins, inflated lungs, the stomach in peristalsis. To see these in action meant overcoming the most insidious epistemological obstacles inherent in the method of dissection of the deceased and to which Aristotle had already alluded.[56] The same argument was also used by Empiricist

50. "Igitur ii qui rationalem medicinam profitentur haec necessaria esse proponunt: abditarum et morbos continentium causarum notitiam, deinde evidentium; post haec etiam naturalium actionum; novissime partium interiorum" (Celsus, *De medicina, pr.* 13).

51. "Praeter haec, cum in interioribus partibus et dolores et morborum varia genera nascantur, neminem putant his adhibere posse remedia qui ipsas ignoret"; "Neque enim, cum dolor intus incidit, scire quid doleat eum qui qua parte quodque viscus intestinumque sit non cognoverit, neque curari id quod aegrum est posse ab eo qui quid sit ignoret" (ibid, *pr.* 23, 25).

52. "Aptiusque extrinsecus imponi remedia compertis interiorum et sedibus et figuris cognitaque eorum magnitudine" (ibid, *pr.* 26).

53. Ibid, *pr.* 24.

54. Ibid, *pr.* 23.

55. Ibid.

56. Aristotle, *PA,* 641 a 1–5: "No part of a cadaver, for example an eye, a hand, is any longer truly such" since it will no longer be able to fulfill its proper function, "just as it would be impossible to do so for a flute of rock or a painted physician."

physicians, in opposition to the Dogmatists, when they denied the utility of the dissection of cadavers, sustaining that everything changed in man the moment he died. Vivisection, the killing of the subject to be observed with the first incision, was according to the Empiricists a useless operation as far as anatomical knowledge was concerned, and was in addition morally reprehensible.

Celsus's text, in addition to furnishing an explanation for the medical assumptions behind dissection and vivisection, also offered certain criteria for the selection of the bodies that would be profaned. Vivisection, and probably also dissection, was carried out on the bodies of condemned men handed over to physicians by Egyptian sovereigns. Their dissection thus became one of the punishments imposed on those who transgressed the norms of behavior sanctioned by the law: vivisection (or dissection) was associated with punishment. At the same time the fact that the body had been obtained through the concession of the sovereign provided the physician who practiced on it with a powerful legitimation. This is the first appearance of the model that would be followed by doctors and European judicial and university authorities in the fifteenth and sixteenth centuries discussed in the second chapter. Its essential characteristics (the opening up of the cadaver as punishment, the legitimization provided by the placet of the authorities, the status of the cadaver marked for dissection) were extremely resistant to the passage of time and to many different scientific, political, religious, and cultural contexts.

The royal concession of the bodies of prisoners for vivisection, as related by Celsus, constituted a second element, alongside the assumptions of the Hellenistic school of Dogmatist medicine, that made the anatomical investigation of the human body conceivable and then possible. This shows how important it was to have a political and cultural climate sympathetic to scientific research so that the practice could be put into effect, as was the case during the reign of the first two Ptolemys. Even if Dogmatist medical theory had made evident the necessity for anatomical knowledge and had proposed the technique to pursue it, only an appropriate context could permit such a theory to be formulated and translated into practice. Favorable circumstances promoted the development of scientific research in the form of patronage, furnishing the required instruments, space, and financial support. Beginning with Herophilus, thanks to state support, the school of Alexandria began its celebrated history of interpreting the new post-Hippocratic developments. The dissection of the human body, in fact, continued to be employed in Alexandria as a research and instructional tool for a number of centuries. Galen reported this to his reader in 177 A.D.: "It must be your assignment and commitment not to learn the form of each bone only from books, but rather to become an assiduous eyewitness observer of human bones. This is easy enough in Alexandria, since the physicians of the place offer to their students teaching accompanied by visual observation. If for no other reason, you must find a way to get to Alexandria at least for this."[57]

A Paradigm for a Millennium

Looked at retrospectively, Galenic anatomy on the one hand represented the high point of scientific production on this subject for more than a millennium, but on the other was, or so it seems to me, one of the forces mainly contributing to the abandonment of the practice of dissection in the West until its revival in the thirteenth century. In effect, it brought to a halt the further development of anatomy until the mid-sixteenth century. The present section will concentrate on some of the characteristics of this solid and enduring paradigm that Renaissance anatomists eventually had to confront, and it will trace in broad outlines the transmission of Galenism from antiquity to the modern era. This reconstruction is necessary because fifteenth- and sixteenth-century treatises held up Galenic anatomy as something that constantly had to be contrasted to their own activity; it is also needed to explain the paradox of the stagnation in anatomical study and in the practice of dissection that marked the history of this discipline for so many centuries.

Galen's *Corpus* contains twenty-six writings dealing primarily with anatomy and physiology,[58] and they constantly allude to and enjoin a

57. Galen, *De anat.*, K II 220–21. There is deep disagreement in historiography over the length of time that dissection persisted in Alexandria. Edelstein's hypothesis, "dissection began in the third cent. B.C. in Alexandria, and only in Alexandria was it still possible in the second cent. A.D." ("History," p. 251), is opposed by the more recent opinion of von Staden, in whose view dissection enjoyed a much briefer history that ended with the first two Ptolemys (end of the third, beginning of the second century B.C.) (*Herophilus*, p. 148). It is curious that the proponents of the latter view ignore the evidence provided by Galen as well as that offered by the Empiricist physician Apollonius of Citium, who accuses Hegetor, a Dogmatist anatomist of the first century B.C., also mentioned by Celsus, of having performed dissections on the human body (see H. Schone, *Apollonios von Kitios* [Leipzig, 1896], pp. xxv–xxvi).

58. See the classification in L. Garcia Ballester, *Galeno, en la societad y en la ciencia de su tiempo (c. 130–c. 200 d.-de C.)* (Madrid, 1972), pp. 260–61. For a chronological census of Galen's works, see J. Ilberg, "Über die Schriftstellerei des Klaudius Galenus," *Rheinisches Museum für Philologie* 44 (1889): 207–39; (1892): 489–514; (1896): 165–96; (1897): 591–623. See also, although it considers only Renaissance editions, R. J. Durling, "A Chronological Census of Renaissance Editions and Translations of Galen," *Journal of the Warburg and Courtauld Institutes* 24 (1961): 230–305. The standard edition of Galen's works usually consulted are his *Opera omnia*, ed. K. G. Kühn, 20 vols., (Leipzig, 1822–31) (reprinted Hildesheim, 1965; microfiche ed. Oxford, 1976). This edition also contains works that have

study of the structure of the human body for the practice of a medicine founded on logic and on apodeictic demonstration.[59] From the time that he was a medical student Galen came into contact with the school of Marinus, certainly the liveliest of the late Hellenic period and the most direct in the line deriving from the teaching of Herophilus. In fact, he studied at Pergamum with Satyrus, at Smyrna with Pelops, and at Corinth and Alexandria with Numisianus, all of whom were in some way tied to the old anatomical school of Marinus, who had lived a generation or two earlier and was the author of an anatomical treatise in twenty books summarized and commented on by Galen himself.[60] As a student the latter had the opportunity to learn anatomy, both theoretically as well as through the dissection of animals and, though rarely, of humans.

To appreciate the role played by anatomy in Galenic medical theory, and thus for all of late antiquity, the Middle Ages, and part of the modern era, we need to turn back to the greatest of the anatomical and physiological texts: the *De usu partium,* and the *De anatomicis administrationibus.*[61] Through these two works it is possible to isolate the two principal intersecting and often overlapping arguments—one of a clinical and the other of a primarily philosophical nature—that Galen used to justify the

---

been wrongly attributed to Galen and fails to include texts that instead are truly his. On this issue, see in vol. 20 of the Hildesheim reprint: K. Schubring, "Bemerkungen zu der Galenausgabe von K. G. Kühn und zu ihrem Nachdruck," in Galen, *Opera omnia,* vol. 20, ed. K. G. Kühn (Hildesheim, 1965; 1st. ed. Leipzig, 1822–31), pp. 1–62. Cf. V. Nutton, *K. G. Kühn and His Edition of the Works of Galen* (Oxford, 1976).

59. See M. Vegetti, "Modelli di medicina." In the *De sectis,* a treatise describing contemporary schools of medicine, Galen quarreled with all those who, whether Empiricist or Methodist, considered the in-depth study of anatomy to be useless (Galen, *De sectis,* 5; K I 77 e 9; K I 96). On Galen's anatomy, see also L. Garcia Ballester, *Galeno,* chapter 3 and the bibliography at p. 111, note 1; F. Kudlien, "Anatomie," *RE,* Suppl. XI, 1968, cols 38–48; O. Temkin, *Galenism: Rise and Decline of a Medical Philosophy* (Ithaca and London, 1973), chapter 1, especially pp. 40–43.

60. This is the *Excerpta ex libris anatomicis Marinis,* a lost text that Galen speaks about in his *De libris propriis 3:* in Galen, *Scripta minora,* ed. I. Müller (Leipzig, 1891), 2:104–8, and in *De anat.,* K.II 280. Thanks to Galen, who calls Marinus "the restorer of anatomy," the latter's fame came down to the Renaissance: a number of authors, including Vesalius, rank him as one of the greatest anatomists in the history of medicine even though none of his writings have survived. Galen tells about his teachers and the places where he studied in *De anat.,* K II 217–18.

61. On these works, see the introductions and the bibliographies of the English translations with commentaries: M. T. May, *Galen on the Usefulness of the Parts of the Body* (Ithaca and London, 1968); C. Singer, *Galen on Anatomical Procedures* (London, 1956); D. E. Duckworth, trans., *On Anatomical Procedure: The Later Books* (Cambridge, 1962). The *De anat.* was recently translated into Italian by Ivan Garofalo, who provided it with a brief but highly effective introduction: Galeno, *Procedimenti anatomici,* 3 vols. (Milan, 1991). Cf. R. French and G. E. R. Lloyd, "Lost Greek Fragments of Galen's Anatomical Procedures," *Sudhoff Archiv* 62 (1978): 235–249.

central place of anatomy. Galenic medicine considered itself a rational discipline (*logike*), one certainly based on the observation of pathological phenomena, the effect of drugs, and the evaluation of obvious causes (the same presuppositions as those of the Empiricist school, as we will see later). But the duties of a good doctor did not end in the establishment of experiential data (*peira*). For it to be rational—and for Galen this signified that it had to be real medicine and not charlatanism—it had to be based on both logic and on apodeictic demonstration applied to incessant investigation, not only of obvious causes but also of the obscure origins of phenomena (as the teachings of the Dogmatist school also dictated). This research into the causes of pathological phenomena could not be separated from a detailed acquaintance with the entire structure of the human body. Rational medicine, linking together theoretical considerations and data furnished by experience, had anatomy as its fundamental prerequisite, the sole basis on which this data could be subjected to evaluation and rational development. This assumption was based on the belief that there was an unambiguous relationship between structure and function and that pathological phenomena were nothing other than defects in this relationship.[62] Familiarity with the human body provided the physician with the knowledge to formulate, with the help of logic, scientific opinions on prognosis, diagnosis, and the correct therapy.[63] Nevertheless, Galen was forced, though reluctantly, to admit the clinical inefficacy of this knowledge that could interact so poorly with the pathology of humors.[64] Anatomy, instead, was recognized as more directly indispensable for the practice of surgery.[65]

Thus far, Galenic reasoning in favor of anatomical knowledge for the physician did not differ much from that underlying Hippocratic medicine.[66] Still, Galen's anatomy was comprehensive and detailed, responding

62. See M. Vegetti, "Modelli di medicina," pp. 116–17.

63. "It is appropriate that anatomical theory should be included in books about prognosis, diagnosis and therapy, as Hippocrates had justly done," (Galen, *De anat.*, K II 282).

64. Idem, *De usu*, 17.

65. The early pages in the third book of the *De anatomicis administrationibus* treat precisely the relationship between anatomy and surgery. Detailed knowledge of the limbs (bones, muscles, veins, arteries, nerves) and of all the other external parts of the human body is indicated as the most necessary for persons embarking on their medical apprenticeship. As Galen remarks repeatedly, ignorance of these aspects of anatomy can lead to error in the prognosis or, worse, to irreparable harm to the patient (*De anat.*, K II 340–46). In the second book Galen writes: "If a person does not know the placement of the vital nerves, of the muscles, of the arteries or of the important veins, he is more likely to cripple or kill his patients than to save their lives" (ibid, K II 284).

66. In the writings attributed to the mythical physician of Cos, this, as I have already mentioned, always remains implicit. To explicate it Galen provides a commentary to a non-

to a broad range of queries, from surgical and clinical to philosophical.[67]
The *De anatomicis administrationibus* described not only the parts of
anatomy considered to be most immediately necessary to medicine, but
also all those the knowledge of which had been deemed "excess," suitable
only for providing the accurate responses to questions about nature asked
by the philosophers.[68] The intelligent anatomist, wrote Galen, had to
draw out of his observation matter "useful both for the practice of medicine and, secondly, for the knowledge of nature." In this, the Galenic anatomy enterprise surpassed that of the Alexandrian Dogmatist school. It
went beyond, even in its objectives, the boundaries of medicine proper,
since anatomy was considered no longer within the limits of application
to therapy: "we are compelled to investigate the utility of all the parts,
even if this does not contribute a bit to the diagnosis of the disorder or
to the prognosis of what will follow."[69] With Galen, anatomy became a
laboratory in which everyone, not only the physician, could verify the
workings and purposes of nature. Galen articulated the presence of a
strict teleology through anatomy: "in constructing the parts, It [the Mind
that modeled living beings] has always present before itself their uses for
Its purposes."[70] The study of the body, on this larger view, joined the
anatomy of the philosopher to that of the anatomy of the physician. But
what is most relevant for the present discussion is the fact that this view
led Galen to a comprehensive type of anatomical research freed of the
restrictions imposed by medicine and surgery.

Direct observation by means of dissection, in contrast to an anatomy
based on books, was postulated by Galen as the surest way to learn about

---

existent Hippocratic anatomy (*De anatomia secundum Hippocratem*). Not being willing to
admit that the father of medicine had never written specifically on anatomy, Galen, in his
De anatomicis administrationibus, justifies the absence of a contribution on the subject by
Hippocrates, saying that it would have been useless. In antiquity, Galen maintained, knowledge of anatomy was transmitted from father to son and not only among physicians but in
all educated families, with the result that medicine began to be practiced even by persons
who were not part of the great Aesculapian family (*De anat.*, K II 280–82). This eventually
created a need for books of anatomy.

67. The *De usu partium* was written "no less for the philosopher than for the physician"
(Galen, *De anat.*, K II 291). A passage from this treatise reads: "[Physiology] will thus be
useful not only for the physician, but much more to the philosopher who is endeavoring to
learn about all of nature than to the physician" (K IV 361).

68. See the first section in this chapter, entitled "Physicians and Philosophers Working
on the Discovery of the Body, or the Uses of Anatomy."

69. Galen, *De usu*, K IV 365.

70. Galen, De anat., K II 542. Cf. *De usu*, K IV 360 ff. and M. Vegetti, "Modelli di
medicina," pp. 120–24. The teleologism inherent in nature and in a particular way in the
animal species invests not only their physical structure but also their soul. For Galen, a
further proof demonstrating the axiom of a single molding Mind is shown by the fact that
"every animal has a corporeal structure congruous to the character and the qualities of his
soul (psyche)" (*De anat.*, K II 538).

human structure. During his time in Alexandria he participated in dissections on the human body, and this undoubtedly provided a solid base on which to rest part of his findings and to build an investigative methodology. But elsewhere he rarely had such opportunities. The acquisition of the thorough and relatively detailed knowledge of all the parts of the human body, to which Galen's works bear witness, was made possible by a series of learning and investigative techniques devised to circumvent the extreme difficulty of procuring cadavers suitable for dissection, and even an actual prohibition from handling, dissecting, and observing them.

Galen's writings testify to his assiduous consultation of the anatomical and more generally medical texts of the earlier authors. He continually quoted (often minutely commenting and summarizing them) Hippocrates, Diocles, Herophilus, Erasistratus, and Marinus, but even Aristotle and Plato are the pillars on which the whole structure of Galen's work on the human body rested.[71] Books constituted a first fount of knowledge for him, as for all students of medicine, but their acknowledged authority was not enough. They were complemented by techniques capable of leading to the empirical verification of texts and to more efficient memorization for the student.[72] Two roads were open to the anatomist and to the student who lacked access to human cadavers: the study of deep wounds on living or dead bodies and the dissection of animals.

Although Galen frequently studied wounds, especially on gladiators (proposed by the Empiricist school as adequate for the acquisition of anatomical knowledge), he criticized this method and deemed it inadequate. It did furnish data on some superficial parts of the body, such as the muscles and some of the arteries, bones, and nerves, but it never produced knowledge of the internal organs or a comprehensive topography of the body. Moreover, this means of observation was absolutely useless from the didactic point of view.[73] This "occasional" anatomy, labeled by the Empiricists, "the exploration of wounds," Galen called "utterly futile."[74]

The dissection of animals, instead, was of quite different significance. We know that much of Galenic anatomy was based on the observation of animals, living and dead. "Of all living beings," Galen wrote, "the monkey is the one that most resembles man both in its internal apparatus,

71. "There were not only professional physicians among our predecessors studying anatomy, but also philosophers" (Galen, *De anat.*, K II 280).
72. Ibid, KII 220.
73. "When even those who practiced dissection of those parts with leisure could achieve precise observations, imagine if one could learn from the mere observation of wounds" (ibid, K II 289). On the superficiality of the Empiricists' anatomical teaching, see the entire chapter 3 of the second book (K II 287 ff).
74. See Galen, *De compositione medicamentorum*, K XIII 604–5.

muscles, arteries, veins and nerves, as in the form of the bones."[75] Indeed, like Aristotle before him, he maintained that "there is an exact similarity between the parts of man and those of the monkey."[76] This belief and the consequences derived from it as far as the description of human morphology was concerned provoked the sharp rebukes of anatomists from Vesalius onward.

For Galen it was monkeys, and among them especially those that most resembled man, such as the *Macacus innus*,[77] that the anatomist had to dissect so as to be able, analogically, to infer the structure of the human body. It was also essential that this operation be accompanied by a close reading of an anatomical text to help guide the observation. This was, according to Galen, because "if you trust only what you read, without having first familiarized yourself with the bones of the monkey by observation, you will not be able to identify the human skeleton precisely when you suddenly come upon it, nor will you be able to remember it. In fact, the recollection of perceptible things requires a constant association between the two."[78] If monkeys closely resembling man were not available, it was better to content oneself with other species of monkeys or with other animals such as goats, oxen, dogs, lions, bears and, especially, pigs, than to not practice dissection.[79]

Galen, like Aristotle before him, applied an analogic principle between

75. Galen, *De anat.*, K II 219.
76. Ibid, K II 384.
77. Monkeys with round heads, a shortened jaw, and small canine teeth, a small coccyx so that they walk on two legs and use their rear feet like hands (ibid, K II 219 ff. and 533 ff.).
78. Galen continued: "We should observe each of the parts at length, so that it can be recognized if seen suddenly, preferably on humans, otherwise on animals resembling humans" (ibid, K II 223–24). Here an implicit objection is being voiced against the casual observations proposed by the Empiricist school.
79. The pig, especially, is endowed with a morphology of the internal organs surprisingly similar to man's. It is no accident that it appears in a number of sixteenth-century title pages (figs. 22, 23). On the title page of the Giunti 1596–97 edition of Galen's works a dissection performed by the author himself is depicted (fig. 32). One of the illuminated letters in the *Fabrica* of Vesalius shows cherubs who have performed the same procedure (fig. 33). In 1537, together with Dryander's anatomy, an *Anatomia porci ex traditione Cophonis* saw the light of day, precisely in the name of an internal morphological analogy of pig and man: "Quoniam interiorum membrorum corporis humani compositiones, omnino erant ignotae, placuit veteribus medicis et maxime Galeno, ut per Anatomiam brutorum animalium interiorum membrorum partes manifestarentur. Et cum bruta animalia quaedam, ut simia, in exterioribus nobis inveniantur similia, interiorum partium nulla inveniuntur adeo similia ut porci, et ideo in eis, Anatomiam fieri destinavimus." The *Anatomia porci* appears in J. Dryander, *Anatomiae, hoc est, corporis humani dissectionis pars prior, in qua singula quae ad caput spectant recensentur membra . . . Item anatomia porci, ex traditione Cophonis, infantis, ex Gabriele de Zerbis* (Marburg: E. Cervicornus, 1537). The *Anatomia porci* was written at Salerno in the twelfth century.

Figures 32–33   Figure 32: Galen, *Opera omnia,* edited by Fabio Paolino (Venice: Stamperia Giunta, 1597; Biblioteca Angelica, Rome). Figure 33: Andreas Vesalius, the letter "Q" from *De humani corporis fabrica libri septem* (Basel: Johannes Oporinus, 1543; The Wellcome Center for Medical Science, London). Animal dissection, as an alternative or complement to the dissection of the human body, was practiced in classical antiquity and in the Renaissance. Galen's anatomical knowledge was based principally on the observation of monkeys and pigs. If the former were used to demonstrate, comparatively, the anatomo-physiology of the external parts of the body, of its muscular and bone structures, an ancient tradition deemed the dissection and observation of the pig's internal organs a reasonable alternative to human dissection. In the title page of Galen's *Opera omnia* published by the Giunti in 1597, the author is dissecting a pig in the view of an attentive public of figures—identified by name—from different historical epochs (fig. 32). The dissection of a pig also is the subject of one of the illuminated initials in Vesalius's *Fabrica* (fig. 33).

animal and human morphology to anatomy; at the same time, he adopted the view that there was a correspondence between external appearance and internal structure: "parts that function in the same way and have the same exterior aspect must have the same structure internally. Equally, animals which act similarly and have a similar external aspect have a similar nature in all their interior parts."[80] The principle of Aristotelian epistemology that was founded on the analogy of functions was even applied by Galen to forms, and so became one of homology. Building on this theoretical structure, Galen carried out numerous dissections and vivisections on every sort of animal, always referring his observations back to questions of human morphology and physiology. It was precisely on the basis of such considerations that, a few decades before Galen, Rufus of Ephesus instructed his students on how they might compensate for the prohibition against dissecting the human body: "Listen then and take a look at this slave. First, you should commit to memory that which is superficially visible. Next I shall try to teach you what the inner parts of the body should be called by dissecting some animal that bears a great resemblance to a human being. Even if they are not absolutely comparable, nothing prevents us from demonstrating what is essential in each part."[81]

Galen's anatomical treatises thus demonstrate his use of multiple sources and techniques both for anatomical research and for teaching the discipline: the study of texts, the occasional observation of wounds, and analogical anatomy through the dissection and vivisection of animals. The emphasis, however, is always on visual verification, on demonstration that is contrasted to rhetoric and on the sophistry of texts. If there is a single universal method for the establishment of physiological-anatomical knowledge, which in turn could serve as the foundation of rational medicine, it is dissection: "Whence will this demonstration come? From where else than dissection? . . . . Through the method of demonstration it seems clear that it is more useful to observe those who dissect animals. . . . Everything that falls outside this process is superfluous and extraneous, and thus we are able to differentiate a scientific premise based on demonstration from a dialectic, functional rhetoric."[82]

80. Galen, *De anat.*, K II 536–37.

81. Rufus of Ephesus, "Du nom des parties du corps humains," in *Oeuvres de Rufus d'Ephèse. Textes collactionnés sur les manuscrits, traduits pour la première fois en français,* ed. C. Daremberg and E. Ruelle (Paris, 1879), p. 134. On this passage, see also L. Edelstein, "The Development of Greek Anatomy," *Bulletin of the Institute of the History of Medicine* 3 (1935): 235–48, and, more superficially, J. B. Schultz, *Art and Anatomy in Renaissance Italy* (Ann Arbor, MI, 1982), p. 9.

82. Galen, *De placitis Hippocratis et Platonis*, K V 219. Cf. M. Vegetti, "La scienza ellenistica: problemi di epistemologia storica," in Vegetti, *Tra Edipo*, pp. 160 ff.

In addition to the dissection of animals and the authority of a text, Galen tried to verify the analogy on the human skeleton whenever the occasion presented itself.[83] His works testify to an obsessive and unending search for unburied bodies that would be free of any accusation of his having profaned tombs or having shown contempt for human and religious values concerning the dead. Such searching, which necessarily turned up skeletons rather than intact cadavers, was occasionally successful. Here is how Galen described some of his adventures:

> For my part, I saw many [bones] following the opening of tombs or monuments. Once a river, after flooding a crypt that had been hurriedly built, caused it to burst, and with the force of the current washed away the cadaver with its already putrefied flesh but with the bones still perfectly connected, and carried it down for a furlong. The body was finally deposited in a swampy place with steep banks, and it was possible to view it as it would have been arranged by a physician so as to explain it to a student. Once we also saw the skeleton of a brigand killed by a traveler whom he had attacked, and who none of the inhabitants of the region wanted to bury. In fact, out of hate towards him they rejoiced in the fact that the body had been devoured by birds, which after consuming the flesh for two days, left the skeleton behind for anyone who wanted to study him.[84]

Galen's words, particularly those concerning the second body he claimed to have found, reintroduce the problem of the legality of dissection as a practice while at the same time raising that of the criteria to be adopted for the selection of cadavers to be used for anatomical observation. The common link between the two cases is that both corpses were disinterred. Certainly the cadaver of the outlaw (devoured by birds) raises once more, involuntarily and in a simplified form, that model adopted by the Alexandrian Dogmatist school and by Renaissance anatomists: again, the body is of someone who has lived and died as a transgressor. Its appropriation by the physician and the profane use to which it would be put is thus facilitated. In another passage of the *De anatomicis administrationibus* Galen, censuring those who disregarded "the very beautiful things" that could be learned from the practice of or even attendance at dissections, and who thereby gave an eloquent testimony of their disinterest in learning, summarized the categories of human bodies that could be dissected: the corpses of those condemned to death and or thrown to wild animals; outlaws left unburied in the mountains; and cadavers of foundlings,

83. Galen, *De anat.*, K II 221.
84. Ibid, K II 221.

which were especially valuable for demonstrating the great similarity between human morphology and that of the monkey.[85]

The use by physicians of bodies for medical or anatomical experiments crops up on other occasions in the works of Galen. In the *De antidotis* he recalls that the last king of Pergamum, Attalus III (d. 133 B.C.) and the king of Pontus, Mithridates IV (d. c. 150 B.C.), both interested in the effects of toxic substances, experimented on condemned criminals.[86] In the eighth book of the *De anatomicis administrationibus* Galen includes a close description of an operation performed on a slave belonging to Maryllus, a dramatist of his acquaintance, involving an incision into the breastbone and the exposure of the heart.[87] Speaking again of those who had ignored the dissection of animals and direct observation, Galen mentioned the case of certain physicians who during Marcus Aurelius's war against the Germans, even though they were provided with the possibility of performing dissections legitimately on the dead bodies of "barbarians," because of their lack of familiarity with this type of operation profited no more from these opportunities than a cook might have done, and learned nothing from these cadavers except, perhaps, the location of the viscera.[88]

Contrary to what has been maintained, at least from Vesalius[89] to very recent writers,[90] it appears from these episodes related by Galen that he might have had the opportunity of practicing dissection even after his time in Alexandria. With the exception of the cited anecdotes, in his writings Galen never referred to the occasions during which he had practiced or attended the opening of human cadavers for instructional or demonstrational purposes in the Rome of the Antonines. Despite the absence of information on the matter, there are two reasons for supposing that Galen did actually carry out dissections on the human body—even during his Roman years: first there is his insistence on the need for visual observation and for empirical verification not only on animals but also on the human body to which he returns throughout his writings, and the centrality of dissection in his research methodology; second there are the descriptions of certain types of subjects that could be used for autoptic

85. Ibid, K II 385–86.

86. Galen, *De antidotis*, K XIV 2.

87. Galen, *De anat.*, K II 632–34. Cf. H. von Staden, *Herophilus*, p. 147, nn. 18, 20.

88. Galen, *De compositione medicamentorum*, K XIII 605; Galen, *De anat.*, 385.

89. A. Vesalius, *De humani corporis fabrica*, fol. 3v: " . . . nunquam ipsum resecuissem corpus humanum."

90. To take but one example, L. Edelstein "The History," pp. 256 ff.: "Galen and Rufus speak of dissections although they could no longer perform any," and L. Garcia Ballester, *Galeno*, pp. 85–87.

investigations presented by Galen not in hypothetical terms but as recommendations: criminals, barbarians, enemies, slaves, unburied bodies, bodies torn to pieces by wild animals, and children abandoned to die of exposure. All these are cases of persons excluded from society or those on its fringes, and persons who in different ways occupied a place at the bottom of a moral hierarchy. All were the objects of public scorn. In Galen's writings, as in the evidence provided by Celsus about Alexandria, the status of these bodies would appear to provide a means for the practice of dissection free of the mesh of ethical, religious, or anthropological restrictions, difficulties, or prohibitions that usually prevented it. Of course, it was a limited field of action, so that Galen was compelled to appeal constantly to the observation of animals and to the compilation of anatomical texts based essentially on the analogy between human and animal morphology.

The striking contradiction to be found in Galen's human anatomy—written from the viewpoint of morphological analogy, while allowing a small possibility that he had carried out his observations on the human body—needs further elaboration. The passages in which he refers to the dissection of the human body are always in reference to the teaching of anatomy or to the defective education of physicians: he held that the direct observation of the human body was a useful subsidiary instrument in the teaching of the medical disciplines, which were however essentially based on the study of texts. But Galen's allusions to dissection as an indispensable research technique in rational medicine—and not exclusively as a didactic-demonstrative tool—suggest, instead, that the observation of living and dead animals was judged sufficient for the acquisition of that knowledge required by the clinician.[91] From the time of Galen, in fact even before the Dogmatist physicians recorded by Celsus, dissection and the direct visual observation of the human body in its internal components assumed the prominently (if not exclusively) didactic role that it would assume for much of the modern era. In the course of the centuries

91. This is never enunciated clearly in Galen's works, but it seems to be constantly implied and plausibly explains the contradiction that has been identified. Celsus too, in the final lines of the *Proemium* to the *De medicina*, taking a middle course between the position of the Empiricists and the Dogmatists, gives an indication of the use that he thought should be made of dissection: "Incidere autem vivorum corpora et crudele et supervacuum est, mortuorum discentibus necessarium, nam positum et ordinem nosse debent, quae cadaver melius quam vivus et vulneratus homo repraesentat" (Celsus, *De medicina*, pr. 74). Thus dissection is justified exclusively on the basis of the teaching of anatomy; it permits the demonstration of the individual parts of the human body, and the identification of the position, form, function, and reciprocal relationships of the organs. By seeing, the student would recognize the parts and would remember.

dissection and anatomical demonstration was therefore associated with the reading and commenting of texts, and not reflected on or practiced outside the course of teaching.

This link between dissection and education, and thus between dissections and the accompanying texts, noted in the episodes discussed by Galen and reaffirmed in Renaissance anatomical practice is to be found also in certain intermediate pieces of evidence. Galen's works were the object of incessant commentaries and translations from late antiquity to the modern era. Such Byzantine interpreters as Oribasius, Aetius, Alexander of Tralles, and Paul of Aegina;[92] or Arab physicians and philosophers such as Al-Râzi (Razhes for Latins; ca. 865–925), Alî ibn Abbâs al-Majûsî (Haly Abbas, tenth century), Ibn Sînâ (Avicenna, 965–1037), Abû'l-Qâsim al-Zahrâwi (Albucasis, eleventh century);[93] such translators from Greek into Syriac and Arabic as Hunain ibn Ishaq (Iohannitius) and his nephew Hubaish;[94] and such translators from Greek and Arabic into Latin as Gerardo da Cremona, Constantinus Africanus, Burgundio da Pisa, Peter of Abano, Arnaldo da Villanova, and Niccolò da Reggio,[95] in

92. On the role that they played in the diffusion of Galenism, see M. Neuburger, *Geschichte der Medizin* (Stuttgart, 1906), pp. 104–28. Oribasius, in particular, is the author of a monumental medical encyclopedia based entirely on Galen's treatises and on the writings attributed to Hippocrates: the *Synagoge* in seventy volumes, which has not survived. On the popularity of Oribasius and the Latin translations of his works, see H. Morland, "Die lateinischen Oribasius-Übersetzungen," in *Symbolae Osloenses*, fasc. V (Oslo, 1932). This writer's *Collectiones medicae* are available in CGM, 6.

93. On Arabic medicine, see M. Ullmann, *Die Medizin im Islam* (Leiden, 1970) and *Islamic medicine* (Edinburgh, 1978, [1st ed. [German] Leiden, 1970]). Cf. S. H. Nasr, *Science and Civilization in Islam* (New York, 1968), especially chapter 8, and D. Campbell, *Arabian Medicine and Its Influence on the Middle Ages* (London, 1926).

94. On the Arabic translations, see M. Steinschneider, *Die arabischen Übersetzungen aus dem griechischen* (Graz, 1960). On Hunain, see M. Meyerhof, "New Light on Hunain ibn Ishaq and His Period," *Isis* 14 (1926): 685–724 and G. Bergstrasser, *Neue Materialen zu Hunain ibn Ishaq's Galen-Bibliographie* (Leipzig, 1932). Hunain describes the techniques of translation and the philological problems encountered in the process in a text in which he gives an account of Galen's known works (he counts 129), of the Greek manuscripts then available, and of the writings that he translated. This is entitled "Memorial on Galen's books that we have translated and on some that we have not" (Hunain Ibn Ishaq, *Über die syrischen und arabischen Galen Übersetzungen*, ed. G. Bergstrasser [Leipzig, 1925]). Cf. J. Vernet, *La cultura hispano–árabe en Oriente y Occidente* (Madrid, 1978) (French translation, Paris, 1985, p. 175). For the translations of Galen into Arabic, see also R. Walzer, "Djâlinus," in *Encyclopédie de l'Islam*, 2:413–14.

95. For a general overview of the problem, see D.C. Lindberg, "The Transmission of Greek and Arabic Learning to the West," in *Science in the Middle Ages,* ed. D.C. Lindberg (Chicago and London, 1978): 52–90, and the brief but highly useful synthesis of medieval medicine by H. Talbot, "Medicine," in *Science in the Middle Ages,* ed. D.C. Lindberg (Chicago and London, 1978): 391–428. See also G. Sarton, *Introduction to the History of Science* (Washington, D.C. 1927–48), II, t. 1:342–48 and III, t. 1:447–48, and J. Agrimi and C. Crisciani, *Edocere medicos: Medicina scolastica nei secoli xiii–xv* (Naples, 1988) (and

varying measure and in different ways contributed between the fifth and thirteenth centuries to the diffusion of Galenic medicine and anatomy, to the transmission of his texts, in fact (and this is of paramount interest here), to keeping alive and consolidating the anatomical paradigm set out in the *De anatomicis administrationibus* and in the *De usu partium*. During this long period, works produced by physicians, biologists, and commentators reveal an unconditional faith in Galen's anatomical descriptions. Even in the Middle Ages they served to maintain a basically unchanged clinical and etiological paradigm and in fact continued to exceed the limited questions being asked of medical practice. This explains, at least partially, the tendency on the part of numerous writers—from Hunain to Avicenna and Mondino, down to the authors of the high Renaissance—to assemble compendia, summaries, and epitomes synthesizing Galen's anatomical works in order to eliminate what appeared to them to be an "excess" of knowledge.

The circulation and adoption of the anatomical texts produced by Galen and his followers was linked to medical practice and particularly to the teaching of the art in the schools during the Middle Ages. If these schools served to strengthen the hold of Galenism they nevertheless fostered a continued, even if often superficial, interest in anatomy and contributed to the survival of those didactic practices Galen had described and promoted for the education of the physician. Among such practices was human dissection.

In a brief but important article Lawrence Bliquez and Alexander Kazhdan have adduced four pieces of evidence that, contrary to what has long been maintained, attest to the practice of dissection in Byzantium between the fourth and the twelfth century.[96] The first document is a *Commentarius Hexaemeron* incorrectly attributed to Eustathius, bishop of Antioch, now remembered as one of the participants at the Council of

---

especially pp. 11–20 for translations as the basis for medical instruction in the Early Middle Ages). For the translations from the Arabic of Arabic texts (the *Canon* of Avicenna) and Greek texts (many writings from the *Corpus Hippocraticum*) by Gerardo da Cremona, see I. Opelt, "Zur Übersetzungstechnik des Gerhard von Cremona," *Glotta* (1960): 135–70. On Constantine and for a listing of his translations, see M. Steinschneider, "Constantinus Africanus und seine arabischen Quellen," *Archiv für pathologische Anatomie* 37 (1866): 351–410; R. Creutz, "Der Arzt Constantinus Afrikanus von Montekassino," *Studien und Mitteilungen des Benediktiner Ordnung* 47 (1929): 1–44; H. Schipperges, *Die Assimilation der arabischen Medizin durch das lateinische Mittelalter* (Wiesbaden, 1964), especially pp. 27–48. For Peter of Abano, see L. Thorndike, "Translations of Works of Galen from the Greek by Pietro d'Abano," *Isis* 30 (1942): 649–53, and, more recently, M. T. D'Alverny, "Pietro d'Abano traducteur de Galien," *Medioevo* 11 (1985): 19–64.

96. L. J. Bliquez and A. Kazhdan, "Four Testimonia to Human Dissection in Byzantine Times," *Bulletin of the History of Medicine* 58 (1984): 554–57.

Nicaea (325 A.D.).[97] The text speaks of the dissections carried out by physicians on the bodies of men condemned to death as a practice that allows the discovery of things of use to all humanity. Theophanes the Confessor relates, in his chronicle written between 810 and 814, the case of one Christianus, an apostate and chief of a robber band known as the Scamari, who in Constantinople in 756 first had his hands and feet cut off and then was turned over to doctors who vivisected him "to learn the structure of the human body."[98] Simeon the New Theologian, a mystical monk who lived between the tenth and eleventh centuries, when comparing the physician of the soul to the physician of the body in one of his ethical treatises, mentions dissection as a means to acquire knowledge about the concealed parts of the body.[99] The last piece of Byzantine evidence, which dates from 1155, is of the same type. George Tornikes, metropolitan of Ephesus, in the course of praising Anna Comnena, harkening back to a passage in the *De anatomicis administrationibus,* compares the perspicacity of Emperor Alexius I's daughter to that of those sagacious physicians who, rather than syllogism, preferred to acquire accurate observation of the human body through the practice of dissection.[100]

Of course this evidence is slight, but each piece would seem to allude to a rather widespread use of dissection. How can this persistence be explained? The role of a tolerant eastern church in this matter cannot explain it adequately. Examining these sources makes one simple fact instantly stand out. The purpose behind the opening of the cadavers (the search for the causes of illness and the acquisition of knowledge about the structure of the body and the position, form, consistency, and reciprocal relations of the internal parts so as to achieve better results when curing disease and healing wounds), as well as the way in which it was carried out (the utilization of the bodies of men condemned to death or on the margins of society, the preliminary incision made from the sternum to the pubes, the justification adopted for the sacrifice of a despised corpse—that of the common good) are traceable to models of behavior and discourse proper to physicians and philosophers of earlier centuries. Dissection appeared dependent on and tied to declarations and practices forged by an older epistemology, one that had been transmitted through texts (whether original, commented on, or summarized) and through an ancient tradition perpetuated in Byzantine schools of medicine.

The history of the Salerno medical school seems highly relevant when

97. *Comm. Hexaemeron, PG,* XVIII, col. 788.

98. Theophanes, *Chronographia,* ed. C. de Boor (Leipzig, 1883), p. 436.

99. J. Darrouzès, ed., Symeon le Nouveau Théologien, *Traités théologiques et éthiques,* 2 vols. (Paris, 1966–67), II:138–40.

100. G. Tornikes, *Lettres et Discours,* ed. J. Darrouzès (Paris, 1970), p. 225.

one considers questions on the relationship between textual tradition, teaching institutions, and anatomical practice. Founded at the end of the tenth century, the school made particularly prominent the teaching of the clinical and therapeutic aspects of medicine.[101] A turning point in the history of this institution coincided with the circulation between the twelfth and thirteenth centuries, also in Italy, of a large number of writings by Galen and Hippocrates translated into Latin by several polyglot intellectuals, most prominently, Constantinus Africanus. The inclusion of these texts into the thoroughly empirical program of the school brought about a radical revision of its teaching mission, adding the study of the texts of the ancient *auctores* to the mastery of professional practice, and turning instruction increasingly toward the goal of a general cultural education of the physician.[102] A law promulgated by Frederick II in 1241, included as rubric 46 in the third book of the collection of Melfi's new statutes, established a series of norms pertaining to the university education of the physician that mark this change in climate.[103] As a preliminary it prescribed three years of logic for the student, to be followed, for the purpose of obtaining a license, with five years devoted to medical subjects, including a year of surgery. Moreover, the law required that the official texts to be used in lessons by the *magistri*, both for theoretical medicine as well as for its practical aspect, should be the authentic ones by Hippocrates and Galen, thereby excluding apocryphal texts, epitomes, and later commentaries.[104] One of these norms required that the surgeon who desired to practice his profession had to present a certificate drawn up by his professors of medicine, confirming that he had studied surgery for at least one year and, "especially, the anatomy of the human body in the schools."[105] This statement caused not a few misunderstandings in the historiography of anatomy. It was believed that this was the first evidence of dissection being carried out on human bodies in the Christian West,[106]

---

101. On this school, see the fundamental study by P. O. Kristeller, "The School of Salerno: Its Development and Its Contribution to the History of Learning," *Bulletin of the History of Medicine* 17 (1945): 495–551. (Italian trans: "La Scuola di Salerno: Il suo sviluppo e il suo contributo alla storia della scienza," in P. O. Kristeller, *Studi sulla scuola medica salernitana* [Naples, 1986], pp. 11–96, with a listing and discussion of the essential bibliography on the subject).

102. P. O. Kristeller, "La Scuola," pp. 62 ff.

103. J.-L.-A. Huillard-Breholles, *Historia diplomatica Friderici Secundi* (Paris, 1854), pt. 1, 4:235–37.

104. "Magistri vero infra istud quinquennium libros authenticos tam Hippocratis quam Galeni in scholis doceant, tam in theorica quam in practica medicine" (ibid, p. 236).

105. P. O. Kristeller, "La Scuola," pp. 66–68.

106. For example, A. Corradi, "Dello studio e dell'insegnamento dell'anatomia in Italia nel medioevo e in parte nel Cinquecento," *Rendiconti del Regio Istituto Lombardo,* 2d. ser, fasc. XV (1873): 632–49; A. D. White, *The Warfare for Science* (New York, 1876) (Italian

whereas in fact the norm simply stipulated that a student should learn about the anatomy of the human body in class and that this could be accomplished through the study of texts and the dissection of animals, primarily pigs.[107] There is definite testimony about such occurrences at least from the first half of the twelfth century.[108]

Anatomical research between the second and the thirteenth century definitely lacked originality. One need only recall Vesalius's sharp rebuke regarding the literary output of the period: "I review them all (with their indulgence), all who have followed him [Galen], in which group I have up to now been able to read Oribasius, Theophilus, the Arabs, and all of our own authors: if perchance they report something worth reading, it has been borrowed from Galen. . . . They have shamefully reduced Galen to trumpery compendia, yet with these they have never grasped his thoughts, deviating by not even a fingernail's breadth."[109] Even though the *De Anatomicis administrationibus* appears to have had only limited circulation during these centuries, the works of some Byzantine commentators and Arab physicians (especially Avicenna's *Canon of Medicine* and Galen's other lesser anatomical writings, as well as his *De usu partium),* made up for the shortage. They contributed decisively to the transmission of knowledge from antiquity to the modern era about the human and animal body, and of investigative techniques, pedagogical methods, and the epistemological principles derived from Galenic anatomy. A number of insti-

---

translation, *Storia della lotta della scienza con la teologia nella cristianità* [Turin, 1902], p. 400); M. Roth, *Andreas Vesalius Bruxellensis* (Berlin, 1892); J. J. Walsh, *The Popes and Science* (New York, 1911), pp. 63–64. A. M. Lassek (*Human Dissection: Its Drama and Struggle* [Springfield, IL., 1958], p. 62) actually states that "one human subject was allotted every five years," without providing any documentary evidence.

107. See the *Anatomia magistri Nicolai,* in G. W. Corner, *Anatomical Texts of the Early Middle Ages* (Washington, D.C., 1927), p. 67, and the *Anathomia porci* (cited at note 79) attributed to Cophon. It seems plausible that it was written by someone who transcribed Cophon's oral teaching, as occurred with the *Ars medicandi,* another work ascribed to this Salernitan master.

108. This is attested to in one of the anatomical treatises circulating in the school, which recalls the dissections performed by the instructor Matteo Plateario: "qui . . . in anathomiae lectione evidenter sub sociorum oculis monstravit nil inferens ficticii nisi quae oculis propriis ipse vidi et quae probabilibus rationibus et auctoritate sunt munita veterum," quoted from K. Sudhoff, "Die vierte Salernitaner Anatomie," *Archiv für Geschichte der Medizin* 20 (1928): 40. Cf. P. O. Kristeller, "La Scuola," pp. 38–39, 68, and T. Meyer-Steinegg and K. Sudhoff, *Geschichte der Medizin im Überblick* (Jena, 1928), p. 193.

109. A. Vesalius, *De humani corporis fabrica,* fols. 3r–v. Theophilus Protospartarius is the author of one of the most important Byzantine anatomical treatises, the *De corporis humani fabrica libri quinque,* wholly founded on Galenic anatomy (especially the *De usu partium*) and devoid of any originality. It never questions the issue of empirical verification. For the *De corporis humani,* I have used the Latin translation by Iunio Paulo Crasso published in Venice in 1537.

tutions from late antiquity and the medieval period (demonstrated by the Byzantine cases mentioned above and by the medical school of Salerno) served as transmitters and veritable embodiments of Galenism. The history of the Salerno faculty, especially, is particularly significant, since it not only experienced a notable renewal of its teaching at a time when Galenic texts began to circulate, but in 1241, Frederick II himself regulated the education of the physician on the basis of certain Galenic fundamental principles. These included the introduction of logic in the general intellectual formation of the physician, who was no longer considered a mere practitioner but also a theoretician, and the obligation to study anatomy and to attend the dissections of animals for those intending to practice surgery.[110] There was no important development in the study of anatomy after Galen (with the exception of the discovery of lesser circulation, which nevertheless remained ignored for centuries.)[111] This may have been due to the thorough treatment of the subject contained in his works, and also to the growing tendency during this period to relegate anatomy to the domain of surgery, for which, even according to Galen, only a somewhat restricted spectrum of morphological knowledge limited to myology and osteology was required.

Anatomical dissection reflected these conventions. We lack adequate documentation and evidence about the ways it was carried out in the eastern empire, but it is certain that it did not play an essential role in medical teaching and research, given the mediocre results found in Byzantine anatomical works and the silence of the medical texts on this subject. The fact remains that from the third century B.C. to the twelfth century A.D. more or less without a break (but we do not know with what frequency), physicians and surgeons did open up cadavers. There was a continuity not only in the succession of texts but also in places, traditions, and institutions, that made dissection practicable.

The diffusion of medical books and the institutionalizing of the education of physicians were not, however, enough to ensure that human dissec-

---

110. This decree constitutes the first institutional testimonial to the association of surgery and anatomy that would be confirmed by the uniting of these two disciplines in a single chair in western universities in the fifteenth and sixteenth centuries.

111. Ibn al-Nafís (687–1288) was a physician of Damascus who in a commentary on Avicenna's anatomy (*Kitâb charh tachrîh Ibn Sînâ*) provides a description of the lesser circulation, the discovery of which had been attributed, until recent years, to Michael Servetus. This was demonstrated in a thesis presented at the University of Freiburg in 1924 by Muhyî al-Dîn al-Tatâwi (see Vernet, *La cultura*, pp. 268–69). Cf. M. Meyerhof, "Ibn al-Nafís (13th century) and His History of Lesser Circulation," *Isis* 23 (1935): 100–20; A. K. Chehade, *Ibn al-Nafís et la découverte de la circulation pulmonaire* (Damascus, 1955). On possible borrowings, see O. Temkin, "Was Servetus Influenced by Ibn al-Nafís?," *Bulletin of the History of Medicine* 12 (1940): 731–34.

tion, as practiced in the school of Herophilus and enjoined by Galen, would be accepted in the West. This is also true of the other technical, epistemological, and subject-oriented aspects of the ancient anatomical tradition. For dissection to begin to be practiced in the West, dignity had to be restored to the study and teaching of anatomy, which had to be recognized for its epistemological and didactic potential and not for its limited applications in the etiological and therapeutic field. Following the example of Galen, anatomical knowledge needed to be legitimized once more, possible applications notwithstanding. But two obstacles had to be surmounted. First, it was necessary to accept the fact that Galen could have erred in some of his descriptions, which then cast doubt on the principle of *auctoritas;* second, and above all, the prevailing horror associated with the handling and profanation of dead bodies, which hung over the practice of dissection, had to be overcome.

UNEASE, DISGUST, CONTEMPT: ARISTOTLE, THE EMPIRICISTS, AND
CHRISTIANS ON THE DISSECTION OF THE HUMAN BODY

In order to confront the moral and religious obstacles standing in the way of dissection one has to return to the origins of anatomy. It is here that those sentiments affecting the procedures of dissection in succeeding centuries reveal themselves; it is here that certain modes of behavior on the part of Renaissance university institutions originate; above all, it is here, in the very beginning, that the doubts, wavering, condemnations, and fears opposing the direct observation of the cadaver until the fourteenth century are formulated and take root.

One cannot talk strictly of a moral and religious interdiction of dissection before it began to be considered an imaginable and feasible practice, that is, before those episodes known to have occurred in Alexandria. The problem of prohibition only appears when the opening of the human body becomes possible according to a medical, philosophical, or broadly cultural paradigm that underwrites the necessity of this technique.

And yet a statement in Aristotle's *De partibus animalium* has been repeatedly quoted to prove the effective existence of a prohibition against the observation of the human body and of dissection before the third century: "it is not in fact without great disgust that we see what composes the human species: blood, flesh, bones, veins and similar parts." [112] These words, repeated by all those who accepted the thesis of a pre-Hellenistic proscription of dissection, leave no doubt that the revulsion generated by the sight of the interior of the human body could have been a factor,

112. Aristotle, *PA*, 645 a 29–30.

together with others of a purely religious character, inhibiting the practice of human dissection.[113] However, these few lines of Aristotle's were preceded by another sentence, frequently omitted by historians of early science: "If someone should consider unworthy the observation of other animals, similarly he should also regard that of his own species; it is not in fact without great disgust that we see what composes the human being: blood, flesh, bones, veins and similar parts."[114] With this, Aristotle's statement now takes on quite a different significance. These "parts" in man, might be encountered under the most disparate circumstances: wounds, bloody deaths, torture, surgical procedures. The examination of animal parts, repeatedly performed by Aristotle, was not to be considered less worthy than that performed on the human body, and was implicitly mentioned as a permissible and desirable operation. Needless to say, in this context the observation of the human body could be effected even without dissection. Aristotle was seeking to justify *ex post facto* his study of animal parts, alluding to its utility, without mentioning the technique that was or should be used. To be sure "the disgust" remains. But in my view it would be a misconstruction to attribute to the word the significance of a taboo, the burden of a prohibition against the dissection of the human body or of evidence of an insurmountable barrier to the pursuit of anatomical knowledge.

The situation appears to be quite different at the time of and following the implementation of the research method established by the Dogmatist school of Herophilus. This new approach was opposed by the Empiricist faction, which developed out of a schism provoked by students of Herophilus within the school of Alexandria itself.[115] Physicians who considered themselves members of this school appear to have been profoundly influenced by skeptical philosophy, whose Pyrrhonist orientation had a considerable impact on diagnostics and etiology, and predominantly on therapeutics.[116] The Empiricists, as their very name suggests, based all medical practice on experience (*peira*), on the study of the course of disease and on the effect of healing measures (*historia*). To the evidence of those facts observed and experimented through *peira* they juxtaposed

---

113. G. E. R. Lloyd, *Magic*, p. 160 and note 184.

114. Aristotle, *PA*, 645 a 27 ff.

115. Philinus of Cos, a student of Herophilus according to Galen, was probably its founder between the end of the third and the beginning of the second century B.C. (see Galen, *Introductio seu medicus*, K XIV 683). On this point, see P. M. Fraser, *Ptolemaic Alexandria*, p. 359 and note 183.

116. P. M. Fraser, *Ptolemaic Alexandria*, pp. 359 ff. Cf. C. A. Viano, "Lo scetticismo antico e la medicina," in *Lo scetticismo antico*, ed. G. Giannantoni (Naples, 1981), pp. 563–656.

speculation (*logos*), judged to be useless if not actually harmful to the practice of medicine,[117] which had to be based solely on experience, on certain and corroborated facts. Vague investigations into natural phenomena, judged obscure and incomprehensible to the human intellect, had to be abandoned.[118] As Celsus had affirmed, further underlining the contrast between Empiricists and Dogmatists, this signified admitting only the inquiry into the evident causes of diseases, and then only those that could be known and verified, as necessary for physicians. The searching into obscure causes and into the natural functions of organs, a useless pursuit given the impenetrability of nature, should be excluded.[119]

It had been precisely the study of obscure causes and natural functions that had driven the Dogmatists, and especially Herophilus and Erasistratus, to undertake thorough anatomical investigations by dissecting dead and living human beings. With the disappearance of these objectives in the medical practice of the Empiricists, the justification for anatomical-pathological or even anatomical-physiological research fell by the wayside. Dissection and vivisection, for differing reasons and under diverse circumstances, were inevitably condemned and eliminated from the practice and teaching of medicine. And it is Celsus, once more, who provides the most complete and detailed testimony of this process.[120] The Empiricists based their objections to vivisection on two related points: first, vivisection, which could have been justified by the opportunity it offered to observe and study living organs and their functions, was considered a useless procedure, since the subject exposed to the incision died the moment the anatomist penetrated at the level of the diaphragm with his knife or higher up in the thorax for the purpose of observing the organs. The

---

117. Celsus, *De medicina, pr.* 36: "Requirere etiam si ratio idem doceat quod experientia an aliud: si idem, supervacuam esse; si aliud, etiam contrariam."

118. Ibid, *pr.* 31: "Cum igitur illa incerta, incomprehensibilis sit, a certis potius et exploratis petendum esse praesidium, id est iis quae experientia in ipsis curationibus docuerit, sicut in ceteris omnibus artibus."

119. Ibid, *pr.* 39: Thus, "Neque quaerendum esse quomodo spiremus, sed quid gravem et tardum spiritum expediat; neque quid venas moveat, sed quid quaeque motus genera significet."

120. Celsus is particularly precise in explaining the teachings of the Empiricists since, in all probability, it was from sources of an Empiricist stamp that he obtained much of the information that he collected on prior medical history. See K. Deichgraber, *Die griechische Empirikerschule: Sammlung der Fragmente und Darstellung der Lehre* (Berlin and Zürich, 1965), pp. 281–88 and H. von Staden, *Herophilus*, p. 145. M. Wellmann ("Erasistratus," in *RE* VI, coll. 335–36) argues that Celsus's source had been Heraclides of Taranto, undoubtedly one of the leading exponents of the Hellenistic Empiricist school. A detailed discussion of the three ancient schools of medicine (the Dogmatist, Empiricist, and Methodist) can be found in Galen, *De sectis* (K I 64–105).

objectives of the operation were thereby negated.[121] This technical evaluation was associated with one of a moral order: vivisection was a cruel procedure in itself and was that much more pernicious since it was useless as a device for the acquisition of knowledge and contrary to medical ethics. "The rest [is] also cruel, that the bowels and vital organs of living humans be cut and that the art that guards human health should cause someone not only death, but death of the most atrocious sort, especially since some of the things sought by means of violence cannot be fully known, while others cannot be known without committing a crime."[122]

The opinion of the Empiricist school on the subject of dissection, still known through the words of Celsus, was no less severe and, as in the case of vivisection, involved elements of both a technical and a moral order: "For this reason, it is not even necessary to carry out dissection on the dead, which is repugnant even if it is not cruel, since in the dead many features appear otherwise [than they do in the living]."[123] The criticism focused again on two factors: first, dissection was not useful for the study of anatomy, since it presented the scientist with the internal parts of a dead body, significantly altered in respect to those of a living being;[124] second, dissection, even if not as cruel as vivisection, was nevertheless still a repugnant practice.

Both arguments, as transmitted to us by Celsus, echo closely passages from Aristotle's zoological works examined earlier. In particular, the reference to the "disgust" that could arise from the sight of some parts of the human body was used here differently from the way Aristotle used it and was proposed explicitly as one obstacle to the practice of dissecting the human body. It was the first formulation in a medical text that alluded specifically to the uneasiness generated by the opening of cadavers.

To evaluate the precise character of this feeling of revulsion, to distinguish it from the one mentioned by Aristotle, additional data on the attitude of Empiricist physicians toward the inner parts of the human body

121. Celsus, *De medicina, pr.* 42: "Nam uterum quidem, qui minus ad rem pertineat, spirante homine posse diduci. Simul atque vero ferrum ad praecordia accessit et discissum transversum saeptum est, quod membrana quadam superiores partes ab inferioribus diducit—*diafragma* Graeci vocant—hominem animam protinus amittere."

122. Ibid, *pr.* 40. Pliny the Elder (*Naturalis Historia*, 29, 18) echoes the deontological condemnation of dissection when he mentions Cato's reproof of certain physicians: "[physicians] Discunt periculis nostris et experimenta per mortes agunt, medicoque tantum hominem occidisse impunitas summa est."

123. Celsus, *De medicina, pr.* 44.

124. In another passage of his *Proemium* to the *De medicina*, Celsus writes: "Neque quicquam esse stultius quam quale quidque vivo homine est tale existimare esse moriente, immo iam mortuo" (*pr.* 42).

is needed. The medical theory of this school had to contain some rudimentary anatomical insights into diagnostic and therapeutic procedures. It thus recommended the occasional observation of the internal organs and tissues of the human body: "If perchance a still breathing man can be examined, this often offers healers an opportunity. Sometimes a gladiator in the arena, or a soldier in combat, or a traveler wounded by bandits is injured in such a way that an internal organ is exposed, and in another individual another organ." [125]

Such random observations offered an opportunity to acquire knowledge about the location of the organs and about their arrangement, form, and other features, and thus contributed to healing instead of merely inflicting cruel measures on living or dead bodies. [126] Empiricist physicians seem not to have experienced the same revulsion when looking inside the body at blood, flesh, and nerves as had the person mentioned by Aristotle in the passage of the *De partibus animalium*. The practice of occasional observation they adopted as a legitimate and desirable strategy in their superficial but necessarily expeditious anatomical inquiry demonstrates clearly that the foulness (*foeditas*) referred to by Celsus had little in common with Aristotelian "disgust," and that it was closely connected to the practice of dissection and thus with opening the body of the deceased.

Consequently, even if there is no mention in Celsus of an explicit prohibition against the use of cadavers for the study of anatomy, it nevertheless expresses unease. Since the evidence of this that he provides differs from the objections raised by the Empiricists, which are of a technico-epistemological character, it would appear not to be an anxiety derived from the teachings of Empiricist medicine or rooted in ideology. These uncertainties spring from the vast and more uncertain terrain of cultural attitudes toward death and the body about which we know little given the paucity of documentation. It seems that there was no formalized prohibition against the opening of bodies based on mysterious anthropological taboos. The Empiricist school coexisted with the Dogmatist in time, space, and sociocultural context, and yet it appears that no interdiction and no taboo had prevented Herophilus and his disciples from doing research. Dissection and vivisection could be performed on the basis of a theoretically based scientific inquiry that carried a humanitarian justification (namely, the sacrifice of a few evil persons to save many good

---

125. Ibid, *pr.* 43.
126. Ibid: "Ita sedem, positum, ordinem, figuram, similiaque alia cognoscere prudentem medicum non caedem sed sanitatem molientem, idque per misericordiam discere quod alii dira crudelitate cognorint."

ones).[127] The operations were accompanied by prudent and shrewd practices (the use of persons condemned to death, or the royal placet, for example) that could safeguard the moral legitimacy both of the operations
and of those performing them. The unease underlying such practices became evident as soon as the theoretical bias of the medical discourse that
had led to the favorable conditions was modified, or rather at the moment
when Empiricist medicine rejected a priori the possibility of acquiring
pathological and physiological information through dissection and asserted its uselessness for therapy.

The opening up of the human body was not the exclusive prerogative
of medicine and biology. Other circumstances (and this was the case particularly in Egypt) could determine when a body would be opened with
the backing of and, even at the request of, the political authority. The
embalming and mummification of corpses for funerary arrangements was
common in Egypt many centuries before dissection and vivisection were
performed by Hellenistic physicians and long before the Empiricists proclaimed them anathema.[128] Although embalming was firmly anchored to
religious belief concerning the afterlife and was carried out with the placet of the civil and religious authorities, it brought contamination and
consequently resulted in the impurity of the person in charge of the opening of the body and of its evisceration. Diodorus Siculus relates that the
paraschistes, that is, those who opened up the cadavers to prepare them
for embalming, were severely maligned and had to flee at the completion
of their work, since the Egyptians held "that anyone who wounded the
body of a like creature (*homophylos*) by applying force against it was
contaminated."[129] The flight or expulsion of the *paraschistes* can be interpreted as a sort of purificatory rite.[130] It is likely that the practices con-

127. Ibid, *pr.* 26: "Neque esse crudele, sicut plerique proponunt [among Rationalist phy-
sicians], hominum nocentium, et horum quoque paucorum, suppliciis remedia populis in-
nocentibus saeculorum omnium quaeri." It is interesting to recall a passage in St. Ambrose's
later *Commentaria in Epistolam ad Romanos* (*PL,* IV, col. 83), extraneous to the world of
medicine, which echoes the above words of Celsus: "Hoc etiam genere antiqui medici in
hominibus morte dignis, vel mortis sententiam consecutis, requirebant quomodo prodessent
vivis, quae in homine latebant, apertis; ut his cognoscerent causas aegritudinis, et poena
morientis proficeret ad salutem viventis."

128. On this issue, see W. R. Dawson, "Making a Mummy," *Journal of Aegyptian Archeology* 13 (1927): 40–49., and J. Scarborough, "Celsus on Human Vivisection in Ptolemaic
Alexandria," *Clio Medica* 11 (1976): 32.

129. Diodorus Siculus, I. 91. 1–4. Even Herodotus (II. 86) offers a detailed account of
embalming, but makes no mention of contamination and filth.

130. In general, on the relationship between embalming and dissection, I have followed
H. von Staden, *Herophilus,* pp. 149–51, and the bibliography cited there.

nected to Egyptian funerary customs furnished Hellenistic physicians with a precedent that, in association with those contextual and epistemological elements discussed earlier, made the dissection of the human body possible. But beyond these conjectures that cannot be verified there is a further interesting fact: the embalming of the human body and the procedures connected to it were unquestionably practiced with the approval or at the request of the political and religious authorities, and were thus justified by a series of beliefs and legitimized by a cultural system that supported its feasibility. This, however, was not enough to prevent the *paraschistes* from contamination through contact with the dead body, evidence despite everything of the persisting unease connected with the opening of the human body and the handling of its internal parts. A similar concern arose over the dissection of cadavers for scientific purposes. The Empiricist physicians were not in fact cautioning against either contact with or the sight of blood and flesh, but rather against the deliberate opening of the body. It was this operation that was defined as foulness (*foeditas*); the risk was one of impurity and contamination.

While the Alexandrian interlude provides the first and only episode of physicians performing vivisections, dissection continued to be considered by some as a practice of use to anatomy. More specifically, such writers as Celsus,[131] Rufus,[132] and Galen refer to the didactic benefits of direct observation and also yearn for the situation prevalent in Ptolemaic Alexandria where physicians could teach and learn about the human body, which yielded its secrets at the very source. Doctors who on the whole recognized the role of anatomical knowledge nevertheless accepted a regimen of behavior that they did not dare to oppose, despite the epistemo-

---

131. "Incidere autem vivorum corpora et crudele et supervacuum est, mortuorum discentibus necessarium, nam positum et ordinem nosse debent, quae cadaver melius quam vivus et vulneratus homo repraesentat. Sed et caetera quae modo in vivis cognosci possunt, in ipsis curationibus vulneratorum, paulo tardius sed aliquanto mitius, usus ipse monstrabit" (*De medicina, pr.* 74–75). In this passage it is also worthwhile noting Celsus's remarks concerning the different techniques for revealing the inner parts of the body, corresponding to the various ends for which they are being exposed: dissection is deemed essential in teaching anatomy, specifically to demonstrate the position and arrangement of the internal organs, other interesting aspects aside; to this is juxtaposed empirical anatomical research that physicians should perform on the bodies of the wounded. It would thus seem that Celsus is following a middle course between the Dogmatists and the Empiricists in regard to their attitude toward anatomical knowledge and related techniques. An awareness of the didactic utility of dissection was undoubtedly one of the decisive factors contributing to its restoration in medieval medicine.

132. "Once upon a time these subjects [the anatomy of the human body] were demonstrated more grandly on humans," a clear allusion to the Alexandrian phase (Rufus of Ephesus, *Du nom des parties,* p. 134).

logical bias underlying their thoughts and actions. At the same time they were aware of and imparted to the readers of their works the advantages that would derive from a direct observation of the cadaver. This reluctant abandonment of dissection needs an explanation. As there is little available documentation to enlighten us about the mysterious and sudden disappearance of human dissection in the West, hypotheses derived from the fragmentary material available are necessary.

Celsus's text, *De medicina,* which deals with the attitude of the Empiricists with regard to anatomy and to the techniques used by the Dogmatists, provides two basic arguments for the prohibition of dissection: foulness and inefficacy. Both cover rather specific semantic fields, one anthropological and religious (foulness, *foeditas*), and the second, which can be called for our purposes technico-epistemological. On the basis of this first approach articulated by the *De medicina,* it is possible to proceed essentially in two directions: the first consists of verifying the existence of unease (or, if you will, of a prohibition), the roots of which are religious or anthropological in character *(foeditas)*; the second, instead, concerns aspects that are more directly epistemological or related to the history of medicine: the utility of dissection in the study of post-Hellenistic medicine and anatomy; the utility of anatomy for medical practice; the constitution of solid, exhaustive, and binding anatomical paradigms such as the Galenic and the corresponding elaboration and refinement of alternative techniques to dissection; the status of the physician; and the relationship to institutions. The factors that emerge as these two approaches are developed over time provide the major response to the question of why anatomical research on the cadaver was abandoned for a millennium. But there is more: independently of context or of period, a number of features affecting this renunciation will disguise themselves in other forms and will also influence behavior (scientific hesitation, formulaic teaching, typographical precautions), as has been observed in the chapter devoted to the Roman case.

Some of the accusations that have come down to us through Celsus's *proemium* were hurled by the Empiricists against vivisection. They subsisted for a long time and discredited both vivisection and dissection, as well as those who performed them. Celsus unwittingly furnished the opponents of such anatomical investigations not only with very specific arguments but with a veritable lexicon of those prohibitions that had evolved around the practice of vivisection. He used such words as execution (*supplicium*), wicked (*scelus*), violence (*violentia*), atrocious death (*mors atrocissima*); the physicians who performed it were described as brigands (*latrocinantes*) and killers (*iugulatores*). But the adjectives that were used most frequently were inhumanly cruel (*crudelis*) and scientifically pointless (*supervacua*). Identical or at least semantically related words cropped up in later texts to designate the "dangerous experiments" of the anatomist.[133]

Among such passages, the one most frequently cited was contained in Tertullian's *De anima,* written at the beginning of the third century A.D., in which he violently attacked one of its leading champions: "The famous Herophilus, the physician, or rather, butcher, who cut up six hundred persons in order to examine nature, who hated humans in order to have knowledge, explored their internal parts—but he probably did not explore all of them clearly, since death itself changes what has been alive, especially a death which is not a simple one but one which is an error in the midst of the artificial process of dissection."[134] Although these words have frequently been interpreted as a disapproval of the vivisections performed by Herophilus, they can just as easily be applied to dissection.[135] Tertullian's source was Soranus of Ephesus, a physician of the Methodist school, who opposed research through dissection or any other sort of surgical intervention, including phlebotomy, which he equated to murder (*iugulatio*),[136] and laryngotomy, which to his thinking had all the characteristics of a criminal act.[137] Tertullian firmly took up the objections raised against the medical followers of Herophilus by the Empiricist school and attacked the practice of dissection, whether of a living or a dead body. He did so from a techno-scientific as well as from a moral-religious point of view. He asserted that the body subjected to anatomical severing cannot provide the physician with any clear data on the human form for two reasons: first of all because (and recalling Aristotelian and then Empiricist reasoning) a living body experiences notable changes at the moment of death; second, because the cruelty of dissection further alters the body's characteristics. Tertullian on this basis reaffirmed the inefficacy (*supervacuitas*) of research that relied on these practices. Moreover, he discredited the anatomist, paradigmatically personified in Herophilus, by stigmatizing him as a butcher (*lanius*) who out of a thirst for knowledge and a

133. This definition of anatomical activity is found on the tomb of Giovan Battista de Ruschis, dated 1653, in the church of San Frediano in Pisa.

134. Tertullian, *De anima,* ed. J. H. Waszink (Amsterdam, 1947), 10, 4. See the commentary at p. 185. This passage is also discussed by H. von Staden, *Herophilus,* pp. 235–36.

135. See H. von Staden, *Herophilus,* pp. 235–36.

136. It is mentioned by Caelius Aurelianus, *Passiones acutes,* 2. 38. 219. See the English translation and the critical edition by I. E. Drabkin: *C. A. On Acute and Chronic Diseases* (Chicago, 1950). On Soranus, see O. Temkin, *Sorano's Gynecology* (Baltimore, 1956).

137. Kind, "Soranos," in *RE* II, 3. 1, col. 1128.

desire to pry open the secrets of nature brutalized man and thereby revealed his hate of him.

The same dual line of argumentation is to be found in the heated attack launched by St. Augustine against the anatomists in his *De civitate Dei* and *De anima et eius origine,* written between the end of the fourth and the early years of the fifth century. In his principal work the bishop of Hippo chiefly emphasizes the moral aspect: "To be sure, some doctors called anatomists with a cruel zeal for science have dissected bodies of dead men and even of men who died while the doctor was cutting and examining them. Thus they have not humanely, but in human flesh, explored every secret place in order to gain new information about such parts and the kind of treatment to employ, and in what place."[138] This was supported in the same work and in the *De anima* with a firm belief in the uselessness of such wanton action on bodily remains, since attempts to shed light on divine secrets, to unveil what is concealed from view, are always in vain.[139]

Along with the hostility directed against the practice of dissection and vivisection, so vehemently expressed by the two church fathers, a chorus of Christian voices was also raised against surgery. Prudentius defined it as "Hippocratic murder,"[140] and some decades later Fulgentius, in his *Mitologicorum libri tres,* wrote: "More cruel than wars is the court of Galen, which is ensconced among all the narrow streets of Alexandria, where many butcher shops of surgeon-executioners are counted among the houses."[141]

The voice of Vindicianus, a North African physician who knew Augustine and was in the service of Valentinianus II, should be added to these denunciations from the early centuries of the first Christian era. His arguments against dissection resemble those of his contemporary: "Our ancestors who practiced medicine in Alexandria—Rufus, of course, and Philip, Lycus, Erasistratus, Pelops, Herophilus, Hippocrates, and Apollonius—found it proper to examine the bodies of the dead in order to

138. Augustine of Hippo, *De civitate Dei contra paganos,* 22: 24. Taken from Saint Augustine, *The City of God Against the Pagans . . . with an English Translation by William M. Green.* The Loeb Classical Library, 417. (Cambridge, MA, and London, 1972), p. 333.

139. Ibid., and *De anima et eius origine,* IV, VI, 7 in *PL,* XLIV: ". . . nulla intrinsecus nostra viscera noverimus, non medicos empiricos, nec anatomicos, nec dogmaticos, nec methodicos, sed hominem scire arbitror neminem." In the *De anima,* IV, II, 3, Augustine, with equal disgust, alludes to the barbarous practice of vivisection.

140. "Horretis omnes hasce carnifucum manus; num mitiores sunt manus medentium, laniena quando saevit Hippocratica?" (Prudentius, *Peristephanon,* 10, 496–98).

141. Fulgentius, *Mitologicorum libri tres,* I: 16–17, in Fabii Planciadis Fulgentii, *Opera,* ed. R. Helm (Leipzig, 1898), p. 9. The observations of both Prudentius and Fulgentius are cited in the commentary to Tertullian, *De anima,* p. 186.

know for what reason and in what manner they died. Humanity itself prohibits doing this, since all things would be manifest and fully open to those conducting the examination." [142]

The close relationship between these texts is obvious even from this hasty review. Although it is not easy to reconstruct the line of development with any precision, it would seem that the original common source is to be found in the arguments advanced against vivisection and dissection by the Empiricists, as transmitted by Celsus. However, instead of attempting a philological and chronological reconstruction of the sources, we should now turn to the models of argumentation and their more strictly lexical aspects.

The two interdependent premises on which dissection was inhibited, both the anthropological-religious one and the epistemological-historical ones expressed in Celsus's *Proemium*, were invoked by Tertullian as well as by Augustine and Vindicianus in their consideration of where the frontiers of the knowable might be drawn. The Aristotelian argument about the uselessness of both dissection and vivisection (because of the mutations of the corpse) as it was filtered through the teaching of the Empiricists was then restated by Tertullian and argued biologically. It became Augustine's and Vindicianus's line of reasoning that all that was invisible, all that God has hidden from human eyes was inscrutable. If this did not necessarily assume the character of a prohibition of the practice of dissection and vivisection, it nevertheless constituted the delimitation of a field, a rather reduced one, within which it was legitimate to conduct anatomical research.

Further and equally significant changes are discernible in the way in which the targets for these writers' invective have also shifted: from vivisection through dissection to phlebotomy. On this point the passage in Tertullian is rather ambiguous: he seems to be inveighing principally against vivisection, but in such a manner as not to exclude dissection. If Augustine, in the De anima (IV, II, 3) condemns the opening of "living

142. Vindicianus, *Gynaecia*, in K. Sudhoff, *Archiv für Geschichte der Medizin* 8 (1915): 417–18 (cited by H. von Staden, *Herophilus*, T64a, p. 189). Cf. K. Deichgraber, "Vindicianus," in *RE*, IX, A1, coll. 29–36. The text in which this passage appears, entitled *Gynaecia* and written in the fourth century, has long been attributed erroneously to Galen. In fact, it appears for the first time, with the title *De natura et ordine cuiuslibet corporis*, in Galen, *Opera* (edited by Rusticus Placentinus) (Pavia: J. de Burgofranco, 1515–16). The *De natura* was reprinted in 1528, 1541–42, 1541–45, 1542, 1549, [1549]–51, 1550, 1556, 1561–62, 1562–63, 1565, 1576–77, 1586, and 1596–97. It was included in all the editions of Galen's *opera omnia* published from 1515 on, with the exception of the 1522 edition prepared by Scipione Ferrari and published in Venice by L. A. Giunta. This erroneous attribution caused numerous misunderstandings concerning Galen's attitude toward dissection during the Renaissance.

men," in the passage of the *De civitate Dei* just examined, he directs his
accusations strictly against those who cut the bodies of dead people into
pieces. The physician Vindicianus, for his part, considering it pointless
now to waste words on an obsolete practice such as vivisection, only
warns against dissection. At the same time another series of pronounce-
ments between the second and fifth centuries, from Soranus to Fulgentius,
reveal a common repugnance toward certain surgical practices. It would
seem that the problem as far as these writers were concerned lay in the
incision, in the profanation of corporeal integrity, in the contact with
blood, and not in a generic handling of cadavers, nor in the homicidal
practice of vivisection.

Finally, it should be stressed that all these invectives were voiced in a
limited vocabulary that partly recalled the anthropological-religious notions expressed in Celsus. The attribution of cruelty (*crudelitas*) to vivisection, omitting dissection, by both the Empiricists and Celsus was used
by Augustine to qualify the diligence (*diligentia*) with which anatomists
approached the dismemberment of cadavers. The word inhumanity (*inhu-
manitas*), which variously occurs in writings by Tertullian, Augustine,
and Vindicianus belongs to the same category.[143] The Alexandrian vivi-
sectionists were dubbed "brigands" and "killers" by Celsus. Herophilus
became a "butcher," which is equated with physician (*medicus*) in Tertullian's text. And *lanius,* butcher, was in different ways at the root of the
various expressions describing the activities of anatomists and surgeons
in passages from Augustine, Prudentius, and Fulgentius.

It is obvious that the word *lanius,* originally used by Tertullian in his
De anima to designate the anatomist, provides the first link in a chain of
subsequent references. The persistence of the term, its successful use until
the Renaissance, is an indication of the special cultural pertinence of the
attribute for anatomists and surgeons. The term *lanius* (butcher), obviously has a highly pejorative meaning for these writers, and the profession
was considered by the Romans to be among the lowest and most sordid,
as is clear in Cicero and Livy.[144] Its essential tasks, the slicing up of the
animal's body, evisceration, the contact with flesh and blood, made the
person practicing the trade loathsome, so that his activity aroused feelings

143. Tertullian: "Herophilus . . . qui hominem odiit"; Augustine, "inhumane"; Vindicianus, "humanitas prohibet."
144. Cicero, *De officiis*, I, 150–51: "Opificesque omnes in sordida arte versantur; nec
enim quicquam ingenuum habere potest officina. Minimeque artes eae probandae, quae
ministrae sunt voluptatum: Cetarii, lanii, coqui, fartores, piscatores. . . ." T. Livy, XXII, 25,
18: "Unus inventus est suasor legis C. Terentius Varro, . . . loco non humili solum sed etiam
sordido ortus. Patrem lanium fuisse ferunt, ipsum institorem mercis, filioque hoc ipso in
servilia eius artis ministeria usum."

of defilement and foulness (*foeditas*).[145] One need only recall what Artemidorus of Daldis wrote in his *Book of Dreams* (second century A.D.)—that merely to dream about being a tanner was a bad omen; he handled dead bodies and usually lived far from the city, or at least outside the city walls.[146] To define an anatomist as *lanius* (butcher) and the practice of dissection as *laniola* emphasized the analogy. But it also meant implicitly accusing the anatomist of treating the human body like that of a beast, adding to the indignity of the practice of human dissection. On the other hand, the act of *laniare* was not the butcher's alone (and by analogy the anatomist's), but also pertained to wild animals. It followed from this that the designation *lanius* carried with it a connotation of bestiality, adding to the sordidness of the profession and consequently also of anatomical practice.

Just because the texts we have examined were produced primarily by Christian writers should not imply that religion lay behind their virulent accusations against the bloody activities of physicians. There is of course a special accent in these texts that distinguishes them from those of a pagan era. However, what seems to be especially noteworthy is the extraordinary consensus between them found in the arguments concerning the prohibition of dissection.

Although references to the cruelty and inefficacy of vivisection and dissection and to the repugnance felt toward such surgical procedures never allude to any formal regulations prohibiting dissection, they are nonetheless very explicit about the moral considerations behind their more or less radical condemnation of these practices. These were buttressed by the argument that there existed a type of unbridgeable frontier of the knowable (to which Vindicianus makes the most direct allusion). It is to be found implicitly in another guise in the objections raised by the Empiricists against the study of anatomy. The terminology alluding to these assumptions seems to indicate that dissection and dissector were viewed as repugnant and defiling. The word. *lanius* and its derivatives were used in

145. Two Roman texts associate *laniatio* and *foeditas*. The first is Seneca's *Hippolytus*, 1246: "Theseu, . . . nunc iusta nato solve et absconde ocius dispersa foede membra laniatu effero"; the second, Tacitus's (*Hist.*, I, 41): "Ceteri crura brachiaque . . . foede laniavere; pleraque vulnera feritate et saevitia trunco iam corpori adiecta." There is certainly also social prejudice against butchers: "Tous ces bouchers se recroutaient principalement dans la classe des affranchis." (See the entry for "Lanius" in the *Dictionnaire des Antiquités Grecques et Romaines,* ed. C. Daremberg (Paris, 1904). Such bias long endured, and significant traces of it persist into the modern era. For an anthropological approach to slaughtering and the butcher shop in the Ancien Régime, see N. Vialles, *Le sang et la chair: Les abatoirs du pays de l'Adour* (Paris, 1987).

146. Artemidorus of Daldis, *Book of Dreams,* I, 53. This concerns the purification of urban terrain, as in the case described earlier of the Egyptian *paraschistes.*

these texts to lead the reader back into a world of filth, contamination, and impurity, as well as one of cruelty and bestiality. The condemnation of dissection (but also of vivisection and of certain aspects of surgery) implied by these terms is not so much aimed at those who trespassed against the prohibition, but reveals instead those qualms surrounding the practice that had now become explicit: the opening up of the body, contact with blood and with death.

While these texts were being produced the Christian cult of saints and relics was coming into evidence in the cemeteries that, in the Roman world, were situated outside the city walls. The growth of this phenomenon "rapidly came to involve the digging up, the moving, the dismemberment—quite apart from much avid touching and kissing—of the bones of the dead. . . ."[147]

These cultic practices, it is easy to imagine, had something in common with dissection from a technical point of view. They shared with it the transgression of certain anthropological codes such as the inviolability of the deceased, the integrity of the body, and avoidance of contamination arising from contact with blood and death. The cult of saints, like anatomy, was based on a system of beliefs (a theory) that determined the adoption of procedures and acts, and these, in turn, clashed with another system, the anthropological, mostly of a hygienic derivation.[148] The beliefs that justified the handling of relics were powerfully backed by eschatological justifications that overcome inhibitions raised by these codes. On the contrary, the arguments adopted by physicians, especially by the Dogmatists, in support of dissection and vivisection, were severely weakened by the criticism of their adversaries, who called these practices useless. Moreover, even Galen and the supporters of dissection considered that this branch of knowledge had very limited therapeutic value. They therefore limited the practice to its utility in teaching so that only those aspects of anatomy required by surgeons were felt to be absolutely necessary. The doubt sown by the Empiricists, the constraints placed around anatomical investigation, and its relative inefficacy in etiology and therapy, served to uphold the condemnation against opening up the human body and its consequent prohibition. In conclusion, medicine had not yet succeeded in providing a sufficiently convincing theory and a paradigm

147. P. Brown, *The Cult of the Saints* (Chicago, 1981), p. 4. See chapters 1 and 2 on these themes. By the same author, see also "Relics and Social Status in the Age of Gregory of Tours," in P. Brown, *Society and the Holy in Late Antiquity* (London, 1982), pp. 222–250.

148. On this point, see M. Douglas, *Purity and Danger: An Analysis of Concepts of Pollution and Taboo* (London, 1966), especially chapters 1, 2, 9. Cf. Douglas's "Pollution," in *International Encyclopedia of the Social Sciences* (New York, 1968), 12:336–42.

to justify the study of anatomy based on the direct observation of the whole human body. The limits of knowledge, mentioned with regard to the Christian texts, could not be pushed forward in such a way as to circumvent moral and anthropological obstacles.

THE REBIRTH OF ANATOMY

About the year 1270 Guglielmo da Saliceto wrote a surgical treatise in Bologna.[149] Although the work never explicitly mentioned dissection, it is frequently described in the historiography of the subject as being based on the direct observation of cadavers.[150] If this is the case it would be the first piece of evidence attesting to the practice to be produced in the Christian West. In fact, Guglielmo could have drawn his insights from his surgical activity, when he treated the deep wounds of his patients: this type of occasional, nonsystematic scrutiny—already mentioned—had been proposed by the Empiricists as an alternative to dissection.

The first unquestioned evidence of a direct observation of the body for anatomical purposes actually dates to 1315. The Bolognese physician Mondino dei Liuzzi relates in his *Anatomia* that he had dissected the bodies of two women, the first in January and another in March.[151] Mondino's text is organized along the lines of an anatomical demonstration: "Which [anatomy] I have proposed to demonstrate, following no high

149. The first printed edition of the *Cyrurgia* appeared at Piacenza [Giovanni Pietro de Ferrati] in 1476 (IGI, n. 8516). I have used the following edition: *Liber magistri Gulielmi placentini de Saliceto in scientia medicinali . . . qui Summa conservationis et curationis appellatur. Item Cyrurgia* (Venice: [Dyonisius Berthocus or Marino Saraceno], 1490) (IGI, n. 8518).

150. See, for example, A. Corradi, "Dello studio," p. 634; J. J. Walsh, "The Popes and the History of Anatomy," *Medical Library and History Journal* 2 (1904): 10–28; J. J. Walsh, *The Popes and Science*, p. 59; C. Singer, "A Study on Early Renaissance Anatomy. With a New Text, the Anathomia of Hieronymo Manfredi (1490), Text Transcribed and Translated by Westland," in *Studies in the History and Method of Science*, vol. 1 (Oxford, 1917), p. 92; J. B. Schultz, *Art and Anatomy*, p. 16; G. Ferrari, "Public Anatomy Lessons and the Carnival: the Anatomy Theatre of Bologna," *Past and Present* 117 (1987): 53; M. Tabanelli, ed., *La chirurgia italiana nell'alto medioevo* (Florence, 1965), especially pp. 742–50, and E. Coturri, *L'insegnamento dell'anatomia nelle università medioevali,* nona conferenza internazionale "Università e società nei secoli xii–xvi" 1979, (Pistoia, 1982), pp. 131–43.

151. "Et per queste quattro razoni quella donna de la qual feci anathomia l'anno mcccxv del mese di gennaro avea la matrice al doppio magiore che quella dela qual feci anathomia nel medesimo anno del mese di marzo" (Mondino dei Liuzzi, *Anatomia*, in John of Ketham, *Fasciculo di medicina*, 1493 ed., fols. g Vr). On Mondino's dissections, see A. M. Lassek, *Human Dissection*, p. 62; J. M. Ball, *Andreas Vesalius, the Reformer of Anatomy* (St. Louis, 1910); C. Singer, *The Evolution of Anatomy* (London, 1925); J. J. Walsh, *The Popes and Science*, p. 50; and the preface to Mondino dei Liuzzi, *Anatomia, riprodotta da un codice bolognese del secolo XIV e volgarizzata nel secolo XV,* ed. L. Sighinolfi (Bologna, 1930).

style, but only that of the manual operation. . . . "[152] In his summary description of the parts of the human body, simulating dissection, Mondino frequently refers to the first *fen* of the first book of Avicenna's *Canon*, and also, but only three or four times, to Aristotle's *De partibus animalium* and to the *On Regimen in Acute Diseases* of the *Corpus Hippocraticum*. However Mondino's *Anatomia* was primarily based, down to the smallest detail, on the only writing by Galen then available that dealt with anatomy and physiology, the *De juvamentis*, mentioned previously, which had begun to circulate in the thirteenth century. Mondino had probably found several opportunities to observe human anatomy at first hand, although it is quite obvious from his treatise that he dealt most often with animals. Although he practiced dissection, in his descriptions he followed Galen and Avicenna, to the point where he referred the reader directly to their texts without even bothering to verify or even simply to demonstrate the truth of their findings.[153] Mondino uses dissection, in fact, to illustrate, to *demonstrate* the content of the text read *ex cathedra:* it is thus strictly a didactic instrument intended to bolster the instructor's words, to reinforce them with a visual image.[154] The anatomist's unconditional acceptance of Avicenna and especially of Galen, and his obtuseness in regard to the cadaver, will characterize the discipline until the time of Vesalius.[155]

The assurance with which Mondino refers to dissection in his *Anatomia* suggests that the practice had achieved a certain standing and that it had become established some time before in university teaching. If ana-

---

152. Mondino dei Liuzzi, *Anatomia*, fol. IIIr. In his description of the human body, Mondino follows the order of the ancient physiological schema that called for the inspection of the *membra genitalia* first, to continue with the *membra naturalia*, with the *membra spiritualia*, concluding with the *membra animata*.

153. For example: "ma quali siano questi tal membri non si può vedere in questa anatomia. Ma bisogna che lo animale si scaccia in acqua piovana: e questo al presente non è necessario. Et se voi cognoscere questi membri legi nela prima fen del primo canone" (ibid, fol. g Vr). Mondino, just to show another obvious instance, follows Galen in the erroneous description of the *rete mirabile* (a thick knot of blood vessels situated at the base of the brain in some animals, but that is nonexistent in humans), and of the heart (described as divided by a porous septum that was supposed to permit the flow of blood from the right part to the left part of the cardiac muscle). Instead, he follows Aristotle, as transmitted through Avicenna, when he asserts that the heart has three ventricles.

154. J. Agrimi and C. Crisciani, *Edocere medicos*, p. 201. Cf. the title page of the 1493 edition of John of Ketham's *Fasciculo di Medicina* (fig. 1) where such a use of dissection is obvious.

155. This problem of the incomprehensability of the evidence offered through observation, and of the coercive force exercised on anatomical knowledge by the ancient authorities, with Galen in first place, occupies a central place in this work. The issue will be developed further in the final chapter.

tomical demonstrations had been taking place even before Mondino, it is possible that they had been performed by Guglielmo da Saliceto, who had been active in Bologna only a generation before. It also seems that other cadavers besides those of the two women were opened and demonstrated upon by Mondino. Guy de Chauliac, a student working in Bologna under the guidance of Niccolò Bertucci, recalls in his *Cyrurgia* an anatomical demonstration conducted by his teacher over four sessions,[156] the same arrangement followed and suggested in Mondino's *Anatomia*. This scansion of the times of dissection, common to both Mondino and Bertucci, dictated by the different rates of decomposition of the body parts, suggests an early, summary formalization of the practice of dissection. Moreover, according to a *Cronaca Persicetana* cited in M. Medici's history of the Bolognese school of anatomy, a woman, a certain Alessandra Giliani, is mentioned as having been Mondino's *sector* for the dissections that he performed in the Bolognese *Studium*.[157] The early diversification in the roles of those participating in the dissection and the formal convergence of Mondino's and Bertucci's anatomies supports the hypothesis that dissection for anatomical purposes had begun to be practiced in Bologna at a date earlier than the one supposed on the basis of Mondino's evidence. Even Guy de Chauliac seems to allude to established practice when he states in his anatomical treatise: "anatomical inquiry is conducted in two ways, first through the knowledge in books, which although it is useful, does not suffice. . . . The other way through experimentation with the bodies of the dead."[158]

Physician-anatomists, at least in Bologna, seem thus to have acquired a certain familiarity with cadavers between the end of the thirteenth and the first half of the fourteenth centuries, enough familiarity, in fact, to risk condemnation for the theft of a body. On November 20, 1319, four medical students, three from Milan and one from Piacenza, followers of a Master Alberto, were prosecuted for having stolen at night from the cemetery of the church of Saint Barnabas the cadaver of a certain Paxius.

156. "Situato corpore mortuo in banco faciebat de ipso quatuor lectiones" (Guy de Chauliac, *Cyrurgia Guidonis de Cauliaco et Cyrurgia Bruni, Theodorici, Rolandi, Rogerii, Bertapaliae, Lanfranci* (Venice: cura Boneti Locatelli, mandato Octaviani Scoti, 1498), fol. 5v. (IGI, n. 4559). These medieval texts are discussed from the point of view of their didactic presuppositions by J. Agrimi and C. Crisciani, *Edocere medicos,* especially pp. 202 ff. and N. Siraisi, *Taddeo Aldarotti and his Pupils* (Princeton, 1981) pp. 58–59.

157. M. Medici, *Compendio storico della scuola anatomica di Bologna* (Bologna, 1857), pp. 20 ff. This information is accepted by J. J. Walsh, *The Popes and Science,* and by A. M. Lassek, *Human Dissection.* The story of Alessandra Giliani, nonetheless, may be a historiographical myth. No evidence of it or of the *Cronaca* is extant before the eighteenth century.

158. Guy de Chauliac, *Cyrurgia,* fol. 5v.

This Paxius had been hanged the previous day at the order of the *podestà* of Bologna and had been buried in the presence of the priest. A servant of Master Alberto was a key witness. He claimed that he had seen the physician in the place where he usually held his lesson in the company of the four suspects and "various others, with razors, knives and other instruments, cutting the body of the aforesaid dead man, and doing other such things as pertained to the medical profession." And further on: "so that Master Alberto taught them to look at the things that there were to see in the man's body." On December 6 the accused declared their innocence and denied everything. The verdict is not recorded.[159]

This extraordinary document lends itself to the following conclusions:

(1) The students were accused of a nocturnal intrusion in the church of Saint Barnabas, "thereby committing a sacrilege and violating a grave situated in a holy place." They were not being accused of having profaned a cadaver when they cut him up into pieces. If that had been the case then Master Alberto himself would have been liable to prosecution, together with all those who had dissected Paxius's remains on November 20. It would appear that the procedure of opening up and performing an anatomy on the cadaver was permissible and legal. One could ask, hypothetically, what position the judge would have taken if the dissection had been performed by someone who was not a physician or a university instructor. In some ways this is an idle question, but it does highlight the professional context in which these procedures could be legally carried out. The dissection of a cadaver was not legitimate in itself, but it was when practiced in a specific place and, especially, by a certain group of persons who could justify their activity in the name of their calling (physicians, medical students): "carrying out what pertained to the medical profession," as the servant affirmed. It is clear that the ambit within which the physician could carry out these activities legitimately is beginning to take shape. The establishment of these parameters became possible only after the acceptance of the principle that under specific circumstances the cadaver might be opened and examined, and also if the practice of dissection was known to have had a history of legitimacy, however recent.

(2) What do the trial records tell us about Paxius? We know only that he was hanged by order of the *podestà* of Bologna on November 19, 1319, buried on the same day, and disinterred "for the purpose of performing an anatomy," during the night between the 19th and the 20th. The following day the body was seen "cut up," and "dissected," in the words of witnesses, in the home of the students. This is rather

159. The trial records pertinent to this affair have been located in the Archivio di Stato, Bologna, transcribed by Ottavio Mazzoni Toselli and published as an appendix to Medici, *Compendio,* and again as an appendix to M. N. Alston, "The Attitude of the Church toward Dissection before 1500," *Bulletin of the History of Medicine* 16 (1944): 233–35.

slim evidence, but we do possess one very important piece of information: the cadaver was not of an ordinary man but belonged to someone who had been condemned to death by hanging. This brings up a further unanswerable question: would the judge have shown the same lack of interest in Paxius's remains if the body had not belonged to a person condemned to death? That he might not have implies that another decisive factor contributing to the legitimacy of the operation was the selection of a cadaver belonging to a certain well-defined sociocultural category, namely that of the condemned person, and this was valid in all contexts, whether ancient or modern.

(3)   In the testimony presented by Master Alberto's servant he does not seem to be surprised, disconcerted, or indignant over what he has seen. It is clearly not the first time that he has chanced upon his master performing a dissection. In fact, he emphasizes the didactic utility of the procedure at least twice. The scene he describes is that of an anatomical exercise. The participants, under the guidance of Master Alberto, included not only the four indicted students, but "several others" who were probably students as well. All of them, armed "with razors, knives, and other instruments," were busy with dissecting the body and cutting away the parts to be examined. This was assuredly not an anatomy lesson similar to those represented on title pages from the end of the fifteenth century; it more closely resembles the instructional model that later would be called a "private," as opposed to a "public," anatomy.[160] There were certainly no public dissections at the University of Bologna in 1319. The first mention of them occurs in the following century. The more informal type of lesson never appeared in the academic calendar and in some universities was fiercely opposed even in later centuries. They presented a problem of a certain significance: how could cadavers be procured for dissection if it was not possible to acquire bodies through the *Studium,* since the course was not part of the official offering? Thus, in the fifteenth and sixteenth centuries, and even more so in the fourteenth (the case of Master Alberto certainly could not have been unique) the students themselves had to cast about for the bodies of recently deceased persons to enhance their education, for which in any case they paid additional fees to their instructor.

Given these facts, it is clear that between the end of the thirteenth century and the first two decades of the fourteenth the first dissections took place in Bologna within the university's medical community. Moreover, these anatomies, although not officially included in the academic curriculum (as revealed by contemporary witnesses), show similarities in the way that they are conducted (the sequence of incisions, for example) that suggest an early, rudimentary standardization of procedures. They also con-

160. On this matter, see chapter 4, pp. 188–194.

tribute some other apparently basic elements refining the definition of the principles that would increasingly regulate the practice of dissection: dissection as the classical authors (Celsus, Rufus, Galen) taught, was a didactic instrument that complemented the study of the anatomical texts and made the acquisition of knowledge of the human body and of all its individual parts possible. It could be useful in the practice of surgery and for the general intellectual formation of the young physician. It had to be performed under the direction of a university instructor, and the cadavers marked for dissection had to belong strictly to condemned persons.

In the course of the fourteenth century dissection was introduced into other universities. Chapter 13, dealing with anatomy, of the 1340 Montpellier University statutes, called for a dissection to take place every two years.[161] In 1376, thanks to a decree of Duke Louis d'Anjou, governor of Languedoc, the dissection of a condemned person's cadaver in that university was approved on an annual basis.[162] In Florence as well, the "Reform" of the statutes of the guild of physicians and apothecaries in 1372 allowed for the practice of anatomical dissection "to the extent that circumstances and fortune permit it in the manner and form used in the universities in accordance with the statutes of the Commune of Florence."[163] In 1391 Juan of Aragon consigned a cadaver every three years for the purpose of anatomical demonstrations to the University of Lerida.[164] In Venice a decree of the Great Council promulgated in May 1368 enjoined the College of Physicians to carry out an annual exercise on a cadaver.[165] The oldest statutes of the University of Bologna (1405) established under

161. "Quia experiencia optima rerum magistra dicitur, statuimus quod, de biennio in biennium ad longius, Cancellarius, una cum Magistro non legente, sacramento sint astricti ut provideant quod fiat anathomia corporalis" (M. Fournier, *Les Statuts et privilèges des universités françaises depuis leur fondation jusqu'en 1789* [Paris, 1890–94], 2:66). Cf. L. Dulieu, *La médecine à Montpellier* (Avignon, 1975), 1:134, and E. Wickersheimer, "Les premières dissections à la faculté de médecine de Paris," *Bulletin de la société d'histoire de Paris et de l'Ile-de-France* 37 (1910): 159–69.

162. This decree was reconfirmed frequently (1377, 1396, 1436, 1484, 1496), evidence of the difficulty in acquiring cadavers experienced by physicians. See L. Dulieu, *La médecine*, 1:131–36.

163. R. Ciasca, *L'arte dei medici e speziali nella storia e nel commercio fiorentino, dal secolo XII al XV* (Florence, 1927), p. 278. Cf. K. Park, *Doctors and Medicine in Early Renaissance Florence* (Princeton, 1985), p. 38.

164. M. Roth, *Andreas Vesalius Bruxellensis* (Berlin, 1892), p. 13, who draws his information from *La Gazette des Hôpitaux*, n. 54 (1881): 430.

165. AS Ven., *Deliberazioni del Maggior Consiglio*, 19 Novella (1350–1358), fols. 114v–115r. Another session, dated August 5, 1370, obliged physicians, and not only surgeons, to participate in the dissections and contribute to the expenses. These documents are cited in G. Ongaro, "La medicina," p. 94.

rubric LXXXVI (*De anothomia quolibet anno fienda*) the norms regulating the practice of public anatomies.[166] The Faculty of Arts at the University of Padua, in statutes revised and published in 1465 and harkening back to older legislation now lost, subsequently recommended anatomical demonstration for the purpose of learning human anatomy and laid down the modalities to be followed and the criteria for the selection of cadavers (foreigners and criminals). This detailed formalization of the anatomical procedure through a series of regulations suggests that dissections had been performed before the date of publication of the statutes. This is in fact documented in a sixteenth-century edition of the treatise *De antidotis* by Leonardo da Bertipaglia, who mentions two dissections in which he had participated in Padua in 1429 and in 1430.[167] Even earlier, in 1404, the Paduan physician Galeazzo Santa Sofia, visiting Vienna in the autumn of that year, had given a public anatomy lesson that was accompanied by a demonstration on the cadaver.[168]

At this point the crucial question is: what made it possible for this practice to become established at that time, first in Bologna, and subsequently in most of the other great European medical faculties? In reply, I could point out that the adoption of dissection in anatomical instruction should not be ascribed to a comprehensive religious and anthropological change in attitude: respect for the dead and revulsion and qualms regarding the opening of cadavers are two factors that supported the strong resistance to change over the centuries. There is an underlying trace of them in every piece of evidence covering the practice of dissection. Nor should its acceptance be ascribed to a new attitude within the medicine of the time, which was still grounded on the humoral paradigm and on the ancient authorities. Mondino performed his dissections, for example, neither taking for granted nor proposing a fresh consideration of the assumptions underlying classical anatomy and medicine. His points of reference while writing the *Anatomia* were exclusively Hippocrates, Aristotle, Avicenna and, principally, Galen.

Instead, it would seem that three conditions may have contributed decisively to the establishment of dissection as a feasible and, later, a legitimate practice in anatomical research and education: the circulation of the

166. *Statuti dell'Università*, pp. 289–90.
167. For more detailed information on anatomy in Padua and on the dissections mentioned by Bertipaglia, see G. Ongaro, "La medicina," pp. 93–96.
168. See A. Corradi, "Dello studio," p. 633. On the first pieces of evidence concerning dissection in the Middle Ages, see also C. Singer, *A Short History of Anatomy and Physiology from the Greeks to Harvey* (New York, 1957); E. Wickersheimer, "Les premières dissections," p. 160; G. Ongaro, "La medicina," pp. 94–96; and M. C. Pouchelle, *Corps et chirurgie à l'apogée du Moyen-Age* (Paris, 1983), especially pp. 137–43.

authoritative Greek and Arab texts, the development of university teaching, and the use of the autopsy.

The translation of the works of Hippocrates, Galen, and the Arab physicians brought to the West a body of knowledge and methodologies affecting both professional and educational practice, including an attention to the study of the structure of the human body. One need only recall how forcefully Galen had stressed this point for the formation of both the physician and the philosopher, despite its limited applicability and its therapeutic inefficacy. For its part, Arabic medicine treated those aspects of anatomy more appropriate to research as secondary to clinical experience. But through classification and revisions of the earlier traditions it also contributed to the part played by anatomy. The chronological convergence of available ancient authoritative texts, which emphasized the cultural importance of knowledge about the parts of the human body and in which both animal and human dissection were discussed, and the revival of this discipline and practice in the West, is certainly not accidental. The history of the Salerno school is strong evidence of the connection. There the impact of the translations on the teaching tradition had the effect of shifting the emphasis from one of strictly clinical and practical/professional instruction to one that included a more complete theoretical education for the student, in which such disciplines as logic and anatomy would have a place. Between the end of the thirteenth and the beginning of the fourteenth centuries there is a rehabilitation of anatomy in response to the message imparted by the ancient authorities. Following the example of Avicenna in his *Canon*, Rhazes in the ninth book of his *al-Mansûr*, and Averroes in the first of his *Colliget*, all of whom had an ample and systematic description of anatomy, a number of European writers introduced some notions of anatomy in their medical texts. This occurred especially in works of surgery, a discipline that had a stronger bias toward the acquisition of direct and palpable data about human structure than did other branches of medieval medicine. One need only think of the works of Henry de Mondeville (d. 1325) and Guy de Chauliac (first half of the fourteenth century) in which the knowledge of human anatomy, entirely mediated by the ancient authorities, is described as being essential for the surgeon. According to Mondeville, surgeons had to link the "method of manual operation" to the teachings acquired through the study of medicine in general and of anatomy in particular,[169]

169. H. de Mondeville, *Cyrurgia*, ed. J.-L. Pagel (Berlin, 1892), p. 132: "Perfectus cyrurgicus sciat medicinam et addiscat ulterius modum manualiter operandi." In another passage Mondeville stresses the importance of seeing others operate: expert surgeons "habent scientiam ex doctrina et ex quod viderunt alios operari" (p. 68). Cf. J. Agrimi and C. Crisciani,

in order to establish an efficient, learned, and rational practice that could be considered scientific, and to avoid being labeled *empiricus* and unskilled (*imperitus*) in an art that was still deemed purely mechanical. This is not to suggest that the study of anatomy, between the end of the thirteenth and the early years of the fourteenth century, was the exclusive preserve of the surgeon. One need only think of Gentile da Foligno (d. 1348), who was well known as a clinician and commentator on the ancient texts and who maintained that a knowledge of anatomy in medicine occupied the same role as the mastery of the alphabet for anyone who wanted to read.[170]

If the rehabilitation of anatomy among the disciplines taught to medical students had to be assumed before dissection could be justified, the educational institutions in which it was practiced, the universities, endowed it with legitimacy and acknowledged its necessity. Dissection, the visual demonstration (*ostendere sensibiliter,* Mondeville wrote) of the anatomical texts read out by the instructors, became a didactic instrument, making the explanation more accessible and facilitating the memorization of the parts being described. The practice that had provoked such qualms among ancients and moderns alike found its only possible context within the university structure.

However, texts that claimed the utility of dissection and institutions that defined the criteria under which it could be performed and guaranteed its legitimacy were not by themselves always able to provide for the opening of cadavers for educational and (even less) research purposes. They only constituted two preliminary conditions. They had to be accompanied by a sort of familiarity with the handling of the dead, permitting the stigma against it to be overcome. I believe that the autopsy, a means of ascertaining the causes of death, provided the important precedent paving the way for anatomical demonstrations on the cadaver.[171] There is evidence of dissection in the service of forensic medicine taking place

Edocere medicos, pp. 189–96 and M. C. Pouchelle, *Corps,* pp. 109, 163. On the relationship between surgery and anatomy at the creation of Medieval universities, see N. Siraisi, *Medieval and Early Renaissance Medicine,* pp. 85–86. On Chauliac and anatomy, see M. Michler, "Guy de Chauliac als Anatom," in *Frühe Anatomie,* ed. R. Herrlinger and F. Kudlien (Stuttgart, 1967), pp. 15–32.

170. Gentile da Foligno, commentary to *Canon,* Bk. I (Pavia, 1510), fol. 15v. Cf. N. Siraisi, *Medieval and Early Renaissance Medicine,* p. 86 and note 11.

171. Dissections for medico-legal purposes are mentioned as decisive precedents for anatomical demonstrations by much of the scholarship on the subject. See, for example, L. Sighinolfi in his introduction to Mondino dei Liuzzi's *Anatomia,* pp. 4 ff.; N. Siraisi, *Medieval and Early Renaissance Medicine,* p. 86; G. Ongaro, "La medicina," pp. 92–94 (with a full bibliography on forensic medicine in the Late Middle Ages).

several decades before it was practiced for anatomical purposes in both Bologna and Venice. Chapter 6 of the civic ordinances of Bologna, promulgated in June 1265, established that two physicians above suspicion (*sine suspitione*) had to examine the wounds of those who had suffered a violent death and to report back on the matter to the court.[172] Similar provisions existed also in Venice (with documentation from 1181 onward) and Padua (from 1276 onward).[173] Although these regulations may imply the use of dissection, it is just as likely that the examinations conducted by the physicians on the slain bodies were carried out without it. Cases of a different type had to present themselves before autopsies in the narrow sense could be performed, with the first piece of evidence dating to 1286, to the Parma chronicle of Salimbene. He gives an account of a curious epidemic during the winter of that year that struck parts of northern Italy, afflicting both humans and chickens: "And a certain physician ordered some [chickens] to be cut open and found an ulceration on their hearts. . . . He also had a dead man opened up and found the same on his heart."[174] The *Conciliator* of Peter of Abano, a physician active in Padua between the end of the thirteenth and the first two decades of the fourteenth century, recalls in the section devoted to poisons (*De venenis*), when discussing mercury, the case of a pharmacist who had mistakenly swallowed a good bit of the metal: "an anatomy was performed on him and coagulated blood was discovered in the area of the heart and similarly in the heart itself. The stomach contained a pound of mercury."[175] Another case of poisoning persuaded a judge to request an autopsy on the cadaver of a nobleman found dead in Bologna in 1302.[176] Even in Paris the documentation for autopsies precedes by some seventy-

172. This document is quoted by L. Sighinolfi in the introduction to Mondino dei Liuzzi, *Anatomia*, p. 4: "Duo medici qui sint sine suspitione et in arte medicandi periti, destinentur ad vulneratum et videant vulnera omnia et sacramento de novo prestito ab eis dicant quot vulnera habet et quot sunt mortifera et quot non mortifera. Et si qua videantur eis facta esse post mortem. Et in hoc quod vulnera sint illata post mortem stetur dicto medicorum." Cf. L. Münster, "La medicina legale in Bologna dai suoi albori alla fine del XIV secolo (nota preventiva)," *Bollettino dell'Accademia Medica F. Pacini* 26 (1955): 257–71, and G. Ortalli, "La perizia medica a Bologna nei secc. XIII e XIV. Normativa e pratica di un istituto giudiziario," *Atti e memorie della Deputazione di Storia Patria per le provincie di Romagna*, n.s. 17–19 (1965–68): 223–59.
173. E. dall'Osso, *L'organizzazione medico legale a Bologna e a Venezia nei secoli XII–XIV* (Cesena, 1956). Cf. G. Ongaro, "La medicina," pp. 92–93.
174. Salimbene de Adam, *Cronica,* ed. F. Bernini (Bari, 1942), 2:357–58. Cf. M. N. Alston, "The Attitude of the Church," p. 226.
175. Peter of Abano, *Conciliator controversiarum, quae inter philosophos et medicos versantur* (Venice, 1565), fol. 263v.
176. The document is quoted by Medici, *Compendio*, p. 39.

odd years the official beginning of anatomical demonstrations at the university.[177]

Anatomical and autoptic dissections had much in common. First of all, they employed the same technique, a similar opening up of the bodily cavity that, in both cases, provoked identical feelings of repulsion and anxiety, and the fear that it could be considered an act of transgression. Moreover, persons of the same professional status performed the two types of dissection. What substantially distinguished the two practices and what made autopsy possible before anatomical dissection were the different purposes behind each procedure. While the study of anatomy did not provide a sufficiently powerful justification for violating the inner secrets of the body and the repose of the deceased, given the uncertain status of the discipline and its untested clinical utility, an autopsy, at least in the cases that have come down to us, was based on solid assumptions. It was performed for the protection of public health (verification of the causes of death during epidemics), to explain suspicious deaths (especially poisonings that could be identified solely through the opening up of the cadaver), to discover the origins of mysterious ailments, and to save countless human lives through the sacrifice of a single body. The latter instance, especially, was part of the research into pathological anatomy in which, as emphasized by the tradition founded by the Dogmatist physicians and defended by Galen, the inspection of the cadaver was considered a possible experimental track for etiological investigations, even if the results had always been relatively modest.

Thanks to the authority of the ancient texts, as soon as the idea was accepted that both the practice and the theory of medicine were inconceivable without a preliminary intellectual grounding that included a knowledge of anatomical structure, and the moment the universities as institutions were able to legitimize dissection, the existence of the autopsy paved the way for a consideration of dissection for didactic ends. Notions of the frailty of the body and the inviolability of the dead were now weakened.

At the same time the rediscovery in the West of human dissection was further justified by the context in which it was practiced: it accompanied

177. The first documented autopsy in Paris—an attempt to discover the ailments that had killed the bishop of Arras so as to be able to care for Dino de Rapundis, chancellor of the Duke of Burgundy, who was showing the same symptoms—dates from 1407. See E. Wickersheimer, "Les premières dissections," p. 6, and M. N. Alston, "The Attitude of the Church," p. 230. The first mention, in E. Wickersheimer, ed., *Commentaires de la Faculté de Médecine de Paris (1395–1516* [Paris, 1915]), dates from the academic year 1477–78 (p. 286).

the reading and commenting of anatomical texts, providing a demonstration that brought about greater clarity and so reinforced through this visual aid the words spoken ex cathedra. Dissection was therefore dictated by a strictly didactic necessity; it was for the sole benefit of students and was not prompted by considerations of research and the acquisition of knowledge. Mondino's *Anatomia* is proof of this. It does not aim primarily at the discovery of the parts of the body but instead at demonstrating the written word. Even Guy de Chauliac provides evidence of this when he places on a par the reading of texts and the dissection of men and animals in the study of anatomy.[178] As for the teaching of anatomy, the objective of the *magistri* and of the universities was limited to passing on to a new generation of physicians knowledge that was both ancient and educationally solid and incontestable. The question of overthrowing these dictates or at least of testing the validity of this body of knowledge never arose. This was a narrow point of view but certainly one of some consequence. Most important, it introduced visualization as an essential educational tool and as an aid to memory in a discipline that for centuries had consisted only of reading, writing, and commentary.

In spite of this array of justifications, the circulation of the ancient texts, and the production of original anatomical treatises, dissection was not to enjoy similar and unconditional acceptance in those universities where the study of anatomy had been introduced. Certain obstacles still blocked its progress. Guido da Vigevano, physician to the king of France, Philip VII, wrote in an anatomical work that he had prepared for his Parisian colleagues in 1345: "Since it is prohibited by the Church to perform anatomies on the human body, and since it is impossible to know the medical art completely, unless one has knowledge of anatomy . . . I shall demonstrate patently and openly the anatomy of the human body, through properly executed illustrations."[179] Here Guido explicitly states that he would use anatomical figures as a substitute for direct observation on the cadaver, a subject that would enjoy great success in the Renaissance and that attests to the extraordinary proliferation of anatomical illustrations that began with the invention of printing.

In a brief treatise inspired by Galen and issued at the same time as Mondino's *Anatomia,* entitled *Anathomia Richardi,* the author writes:

---

178. Guy de Chauliac, *Cyrurgia,* fol. 5v. On the relationship between dissection and teaching in the Middle Ages, see J. Agrimi and C. Crisciani, *Edocere medicos,* pp. 202–3.

179. E. Wickersheimer, "L'anatomie de Guido de Vigevano (1345)," *Archiv für Geschichte der Medizin* 7 (1914): 1. Cf. M. N. Alston, "The Attitude of the Church," p. 225, note 19.

"Now, however, since it is horrible to treat the human body in such a way, modern teachers perform anatomies on brute animals."[180] Guido's *Anatomie* and the *Anathomia Richardi* testify to the uncertain status of dissection: the first alludes to an explicit ecclesiastical prohibition, the second to an ethical judgment that stresses notions of unease and revulsion. A text that superficially does not seem to have much to do with medicine may be useful here: Boniface VIII's bull *Detestandae feritatis* of September 27, 1299. It is clear from it to what degree a hypothetical religious prohibition of anatomy was in fact dictated by ethical considerations, which had an anthropological basis and did not carry any theological connotations.[181] In the bull the pope inveighs with his usual vigor against a custom in use at the death of members of the nobility and the highest reaches of the ecclesiastical hierarchy: at the express wish of the deceased their bodies were dismembered, cut into pieces, and boiled to separate flesh and bones. Two wishes were reflected in this procedure: first, that certain parts of the body (the heart or the intestines, for example) should be buried separately from the rest of the body in order to increase the number of intercessions by burying the remains in several religiously important places; and, second, if death occurred far from one's own lands (in war, for example), and, especially, on unconsecrated soil such as infidel territory, that the body should be boiled in order that the bones be carried back to the place where the deceased wanted to be buried.[182] Boniface threatened to excommunicate anyone who treated or caused to be treated "the bodies of the dead so inhumanely and cruelly." The bull speaks of impiety, savagery, and cruelty in describing these uses. The practice, the bull relates, "fills the minds of the faithful with horror and shocks one's ears." It thus prohibited "that which is rendered not only most abominable in the sight of divine majesty, but is also most vehemently abhorrent to the gaze of humanity."[183] The emphasis placed on

180. *Die Anatomie der Magister Richardus,* ed. J. Florian (Breslau, 1875), pp. 2–3.

181. The text of the bull, in the promulgation of September 27, 1299, is published in Bonifacius VIII, *Les registres de Boniface VIII.* Bibliothèque des Écoles Françaises d'Athènes et de Rome, 2, 4, ed. G. Digard, et al. (Paris, 1904–1939), n. 3409, coll. 576–77. It was issued again on February 18, 1300. Cf. *Extravagantes communes,* III, VI, 1, in *Corpus Iuris Canonici,* ed. E. Friedberg (Leipzig, 1879–81), coll. 1272–73. For a full discussion of the bull, see, especially, A. Paravicini Bagliani, *I testamenti dei Cardinali del Duecento* (Rome, 1980); E. A. R. Brown, "Death and the Human Body in the Later Middle Ages," *Viator* 12 (1981): 221–70; and F. Santi, "Il cadavere e Bonifacio VIII, tra Stefano di Tempier e Avicenna. Intorno a un saggio di Elisabeth Brown," *Studi Medievali,* 3d. ser., 28 (1987): 861–78.

182. For an ample survey of these practices, see E. A. R. Brown, "Death and the Human Body."

183. *Les registres,* coll. 576–77.

the reactions of the faithful and on the fact that the practice undermined fundamental principles of humanity and of respect toward the dead emerges firmly in the bull. Moreover, the document leaves a margin of doubt as to whether the dismemberment of a cadaver was prohibited in any form and in every circumstance. This ambiguity was dissipated in a decretal dated April 19, 1303, where it becomes clear that the *Detestandae feritatis* had not solely forbidden the boiling of cadavers but also opposed the dissection, evisceration, and cremation of the body, as was confirmed, that same year, in a gloss by Cardinal Jean Lemoine.[184]

Clearly the bull and the ensuing commentary discouraged physicians and surgeons from performing dissections for anatomical purposes, as is attested in Guido da Vigevano's *Anatomie* and the *Anatomia Richardi*. In fact, it seems that an apposite papal dispensation was required to practice dissection at this point,[185] although there is no extant documentation to validate the supposition, for example, that Mondino, only a few years after the publication of Boniface's bull, performed the two dissections about which he speaks in his *Anatomia* with prior papal authorization, or that this was also obtained on all the other occasions in which physicians opened up cadavers.[186] It seems possible however to affirm that this bull could have been intended as the first formal attempt at control over anatomy on the part of the church, which was greatly extended in Rome in the course of the sixteenth century. It was seen in that series of measures that affected juridical and religious authorities when the selection of the cadavers of condemned men for consignment to the physicians of the *Studium Urbis* was made, and above all in the approval of the cardinal vicar as papal representative in Rome.

The frequent recurrence in the papal bull to the "humanity" and the

---

184. For the decree of Boniface VIII, see *Les registres,* n. 5218. For the gloss of Jean Lemoine, see *Extravagantes communes omnes . . . cum glossis Ioannis monachii Picardi* (Lyons, 1553), p. 202, gloss to *Extenterant.* For additional details on the subject, see E. A. R. Brown, "Death and the Human Body," pp. 222–23.

185. Even Henry de Mondeville mentioned that the innards can be removed from a corpse for the purpose of embalming him only if "a Romana Ecclesia speciale privilegium habeatur." See E. A. R. Brown, "Death and the Human Body," p. 250. M. C. Pouchelle, *Corps et chirurgie,* especially p. 133, dwells at length on the point that Mondeville could neither dissect nor embalm.

186. A passage in Mondino's *Anathomia,* when speaking of the anatomy of the ear, in numerous editions seems to be referring implicitly to the bull of Boniface VIII: "ossa autem alia quae sunt infra os basilare, non bene ad sensum apparent nisi ossa illa decoquantur, sed propter peccatum dimittere consuevi." The allusion to sin is omitted in some of the Italian translations, such as the anonymous Bolognese one from the end of the fifteenth century, which remained in manuscript and can now be read in Sighinolfi's edition of Mondino dei Liuzzi's *Anatomia* (pp. 184–85).

"piety" that the faithful owe the dead as the essential reason for prohibiting dismemberment shows that its origin lies more in the horror provoked by any practice that infringed on the integrity of the human body
than in theologically motivated reasons, such as, for example, the doctrine of the resurrection of the flesh.[187] Some decades before Boniface's
bull Riccardo and Nicola of Salerno had similarly stressed how dissection
was considered an "inhuman act . . . especially among Catholics."[188] We
find here categories and formulations of prohibition already encountered
in both ancient and Christian texts, which made their qualms about the
opening up of cadavers explicit and stigmatized the practice as "inhuman" and "cruel." It would seem that the sentiment of "humanity" that
these practices were thought to violate crossed over the centuries in the
same way as juridical, institutional, and religious contingencies did and
that they remained one of the most serious obstacles for the anatomist to
overcome before he could open the human body. This continued to be
true at least until dissection ceased to be dependent on the close elucida-
tion of texts and acquired a heuristic value of its own that would find its
justification in the advancement of the medical discipline generally (the
context of research and learning), but also outside it, in the eyes of a
public that viewed the opening up of the human body as an act of transgression.

In spite of the bull of Boniface VIII and of Cardinal Lemoine's gloss,
and in spite of the fact that anatomical demonstrations preserved an eminently didactic character until the first half of the sixteenth century, evidence of dissection turns up with ever greater frequency between the end
of the fourteenth and the early fifteenth century. In these instances the
bodies were opened up in accordance with restrictions that in the course

187. A few years before, in fact, this problem had been much discussed at the University
of Paris, where philosophical questions were raised over the necessity of preserving the integrity of the body even after death in the name of respect for the *cursus naturae*. It appears
that Pope Boniface was kept abreast of this debate and that the bull was promulgated as a
consequence. Cf. especially F. Santi, "Il cadavere," pp. 863 ff., but also E. A. R. Brown,
"Death and the Human Body," pp. 235–46, and A. Paravicini Bagliani, "Storia della scienza
e storia della mentalità. Ruggero Bacone, Bonifacio VIII e la teoria della 'prolungatio vitae,'" in *Aspetti della letteratura latina del secolo XIII. Atti del primo convegno internazionale di studi dell'Associazione per il Medioevo e l'umanesimo latini (AMUL), Perugia 3–5
ottobre 1983*, ed. C. Leonardi and G. Orlandi (Perugia and Florence, 1986), pp. 243–80.

188. On the question of the relationship between dissection and religious prohibitions,
see White, *The Warfare for Science*, but especially, for the wealth of information it contains,
W. Artelt, "Die "ältesten Nachrichten über die Sektion menschlicher Leichen im mittelalter-
lischen Abendland," *Abhandlungen zur Geschichte der Medizin und der Naturwissenschaften* 34 (1940): 2–25 (from which I have taken the quotation). Cf. M. N. Alston, "The Attitude of the Church," and E. A. R. Brown, "Death and the Human body," especially p. 248.

of time assumed a binding regulatory character that was partly inspired by a sentiment of *humanitas* recurring in various documents. These restrictions were the result of philosophical, theological, and juridical values and beliefs; doubt regarding the legitimacy of dissection endured, however: a doubt that had been instilled or at least made explicit, even in its anthropological essence, by Boniface VIII. The ambiguity behind which the ecclesiastical position had entrenched itself was dissolved in April 1482 in a brief sent by the Tribunal of the Penitentiary, at the order of Sixtus IV, to the University of Tübingen, in reply to a request that had been received from the rector, doctors, and students of that institution. The brief granted it permission to receive from the judicial authorities the bodies of such criminals as had been condemned to capital punishment by secular authorities for the purpose of enhancing knowledge and experience of the medical arts through dissection. These bodies could be obtained without petitioning the Holy See, as was the custom, for a dispensation or special license any time an anatomy lesson that entailed a demonstration on the human body was scheduled. This permission was contingent on the normal burial of the bodies of the condemned, namely with all their parts and with appropriate reverence.[189] It is important to underline, however, that this brief did not imply that all dissections were acceptable without regulation or prior apostolic permission. It affirmed instead an unstated, generalized control over every dissection on the part of the religious authority. The church no longer needed to intervene in individual cases but had shown its position on the matter once and for all and had spoken on the general norms that had to be observed. Through this brief the judicial authorities became responsible for the selection of the condemned to be turned over to physicians, while at the same time the responsibility of the judges over the fate of the cadaver remained under the protection deriving from the papal license. The regulations contained in the Roman statutes, as well as those of all the other universities discussed in the previous chapter, hearken back to the brief of Sixtus IV. They, too, give evidence of a significant formal acceptance of the practice, allied to a tacit control on the part of the church over the executors of judicial decisions. The brief, however, maintained that *de*

189. This brief can be read in L. F. Froriep, Über die Anatomischen Anstalten zu Tübingen (Weimar, 1811). Cf. Del Gaizo, *Dell'azione dei Papi sul progresso dell'anatomia e della chirurgia sino al 1600* (Milan, 1893) and M. N. Alston, "The Attitude of the Church," p. 229. It has also been mentioned by Singer, *A Short History of Anatomy*, p. 86, and discussed in another article by J. B. Schultz, "A Fifteenth-Century Papal Brief on Human Dissection," *Medical Heritage* 2 (1986), pp. 50–56. For an English translation of the document, see Schultz, *Art and Anatomy*, p. 221, note 119. He analyzes the contents on pp. 61–63.

facto the body of the condemned person belonged to a terrestrial tribunal that could dispose of it until the moment of burial, when it was materially returned to clerical jurisdiction.

Even though the dispensation had been specifically addressed to the University of Tübingen, Sixtus's brief suggests an explicit approval of dissection by the church. In all probability such dispositions must also have been issued for other universities, such as Bologna and Padua, where dissection on the human body was even more firmly established.[190] The church was thus acceding to the request of the physicians and readers of the *Studia* who, in the name of scientific progress, were asking to perform an operation that anthropologically threatened to violate unwritten but generally accepted codes concerning the management of death and the corpse and that, from a religious perspective, verged on sacrilege.

190. A permit similar to the University of Tübingen's was granted by Sixtus IV in 1480 to the *Ospedale Maggiore* in Milan. See C. F. Biaggi, "Gli studi anatomici all'Ospedale Maggiore nel secolo XV: Leonardo a Milano," *Ospedale Maggiore* 44 (1956): 407–9.

CHAPTER FOUR

Bodies and Texts:
Knowledge and Ritual in Fifteenth- and Sixteenth-Century Anatomy

Two features recur in all examined representations of the anatomical lesson: the open cadaver and a textual authority (a book or a figure reading). The attitude of the physicians and anatomists toward these elements and the manifold uses that derive from them provide the focal point for the history of anatomy between the fifteenth and sixteenth centuries, if not the history of anatomy *tout-court*. As has been suggested, the reversal of the priority between textual authority and a direct observation of the cadaver is the key to understanding Andreas Vesalius's achievement in recasting the anatomical paradigm. This rearrangement is not insignificant, and its practical and theoretical assumptions need to be examined even before his work, which contributed to a redefinition of the attitude of the anatomist toward those tools available for learning, particularly when it was possible to look into the open body while verifying the contents of the canonical anatomy texts empirically. The necessary premises for such a change were, first, the continuous and regularly repeated practice of dissection, especially when it took place outside the limitations imposed by quodlibetarian ceremony and, as a consequence, a familiarity in the handling of the cadaver that might overcome the obstacles of an anthropological order previously discussed. Also necessary was the introduction of doubts about the ancient authorities, which led to the notion that the anatomy of the human body could be rewritten. Some consideration of these factors is required to evaluate the Vesalian shift and the conditions that made it possible.

THE DISMEMBERMENT OF CADAVERS

When Vesalius wrote his *De humani corporis fabrica,* the practice of dissection was already firmly entrenched in many European universities. The spread throughout the continent of this institutionalized and regulated practice strikingly exemplifies the recognition of the didactic and scientific role achieved by anatomy between the end of the fifteenth century and the first half of the sixteenth. The public anatomical demonstrations took place, however, in accordance with a fixed ritual, each step of which was determined by university statutes. In accordance with these customs, as the pre-Vesalian title pages show, access to the cadaver was prohibited to students: they could attend the operations carried out by the *sector* when they would be shown the parts of the body being described in the texts read out loud by the instructor. Basically, dissection was shorn of any research potential, and there was furthermore no possibility of correcting the textual authorities. For the physicians, and even less for the students, the public demonstrations did not bring about familiarity with the cadaver. But, according to Vesalius, of course, only knowledge of anatomy gained by firsthand observation was truly useful to the good physician. In the mid-sixteenth century, he railed, there were still countless "pestilential doctors" who threatened the lives of their patients because they had never dissected "with their own hands."[1]

For dissection to be productive as a teaching tool, and one that furthered knowledge of the human body, it was necessary, first of all, to modify the quodlibetarian model of the anatomy lesson: the physician-*lector* had to come down from his throne and assume the duties of the *sector,* uniting theoretical learning with practical ability; the barrier separating the students from the cadaver also had to be removed. A twofold solution was suggested: the students should be permitted to open the bodies during the public demonstrations and, further, they could organize private anatomies in "seminar" form, with fewer participants.

The first of these solutions was noted when the unusual features of the title page of Winther's translation of the *De anatomicis administrationibus* was examined. Vesalius, in chapter 19 of the sixth book of the *Fabrica,* which is dedicated to the methodology of anatomies to be conducted for the students, emphasized that having them dissect the cadaver them-

---

1. A. Vesalius, *Epitome:* "Serenissimo Principi Philippo . . . : " "Ut pestilentes praeteream medicos, in communis hominum vitae exitium grassantes, qui ne unquam quidem sectioni astitere, quum in corporis cognitione nemo aliquod operaeprecium praestiturus sit, qui non propriis manibus, uti Aegiptorum reges consueverunt, sectiones obierit, et illas perinde ac simplicia medicamina sedulo frequenterque versaverit."

selves was an outstanding technique for the teaching and learning of anatomy. He frequently allowed dissections to be performed even by less experienced and less capable students in front of spectators. At the completion of the procedure, the students would discuss what they had observed in the opened bodies among themselves, deriving enormous profit from their detailed knowledge of this aspect of medicine and, Vesalius specified, also of natural philosophy.[2] Alongside these official occasions, which took place somewhat sporadically, private anatomies were also organized that Vesalius, as well as others, considered pedagogically superior to the public sessions.[3] This type of lecture occurred at the request of a few students, in the privacy of the home of the anatomy teacher, whom they would pay for his labors. The bodies or the body parts used in these sessions were often obtained surreptitiously, contravening the norms laid down in the university statutes. It appears, however, that university authorities not only knew about these private anatomies but even wholly tolerated them.

Such procedures seem to have been a fixture of the history of anatomy from its earliest documented occurrence in the Christian West in the second decade of the fourteenth century. In fact, the oldest known dissections (for example Mondino's in 1315 and Master Alberto's in 1319)[4] took place without any ceremonial restrictions of any kind, and they most likely resembled what would later be called private anatomies. It was not until 1405, in fact, when the first Bolognese university statutes were promulgated, that some restrictions were placed over the private and indiscriminate practice of dissection. The principal concern of chapter 86 of the statutes, dedicated to anatomy, is to regulate a practice that often provoked disorder and protests because of the unrestrained search for cadavers. The regulations limited the number of dissections a student could attend during the entire course of his education to three and estab-

2. A. Vesalius, *De humani corporis fabrica*, p. 547: " . . . multoque minus peritior aliquis, qui ad cuiusque vota sectionem aggrederetur, in promptu semper sit, publicam quoque sectionem a studiosis summopere expetendam duco, ut ipsi vel obiter in ea spectatorum turba edocti, si quando in cadaver incidant, propriis manibus administrationem accurate obeant, et studia invicem conferentes, hanc medicinae seu naturalis philosophiae partem debite amplexentur."

3. Ibid, p. 547: "Quanquam enim privatam et inter paucos exhibitam sectionem, publicae praeferendam nemo ambigat." A request sent by the vice rector of the Faculty of Arts of the University of Padua to the *riformatori* of the *Studium* for permission to conduct private dissections echoes the words of Vesalius: "havendo nui da queste tali Anathomie un certo modo più utilità che dalla pubblica" (*Nuova istanza per ottenere le anatomie private*, December 5, 1590, Biblioteca Universitaria, Padua, *Archivio Antico dell'Università*, vol. 677, fol. 228r, published in G. Sterzi, "Giulio Casserio, anatomico e chirurgo [1552–1616]," *Nuovo Archivio Veneto*, n.s. 10 [1910]: 74–75).

4. See chapter 3, "The Rebirth of Anatomy."

lished that each session should be properly publicized, and they assigned to the physicians responsibility for procuring the cadavers.[5] The private anatomies thus preceded the public ones; the initiatives of a few physicians persuaded the university to institutionalize and control an established practice and to give it a semblance of legitimacy.

The public anatomies, however, did not entirely replace the private ones. The latter reappear in the documentation at a time when ritualized public demonstrations were found lacking within an epistemological context that was undergoing renewal. It was not only Vesalius who encouraged teachers of medicine to arrange supplementary anatomy lectures, and with fewer students. There is widespread evidence of the existence of private anatomical demonstrations that were concurrent with the public ones in Padua, unquestionably the most progressive European center of anatomical study in the sixteenth century. In 1550 the senate was compelled to threaten severe penalties against those who violated graves for the purpose of obtaining cadavers to dissect for "particular anatomies."[6] In 1585 the first requests to drop the prohibition against private dissections appeared.[7] This proscription was implicit in the statutes of the University of Padua, since the rector alone could request the cadaver, and there were limits on the number of annual dissections that could be carried out.[8] But the performance of private anatomies continued: "it is requested," one reads in a petition to the "Prince of Venice" dated February 1590, "that each person before and after the public session be able to participate in private and particular anatomies." The petition had been presented by the vice rector and councilors of the University of Padua in the hope of revoking a "Public Edict" with which "it has been prohibited to all to perform or hold private anatomies," with accompanying "ex-

5. It was the students, however, who had to cover the expenses involved in obtaining the cadavers. See *Statuti*, pp. 289–90.

6. "MDL. nonis sextil. *Gellius a Valle* Vicentinus, Prorector. Ex litter. Ducal. Anatomicum studium in dies magis cum vigeret, non publicae modo, sed privatae quoque exercitationes passim habebantur; quibus si forte cadavera non suppeterent, ne sepultis quidem juventus parcebat. Quapropter Senatus consultum VI. id. febr. factum est gravissimarum poenarum sanctione adversus illos, qui per huiusmodi caussas sepulchra violarent" (J. Facciolati, *Fasti Gymnasii Patavini*, [Padua, 1757] pt. 3, p. 208). In 1601 Casserio was again suspected of having urged students to steal bodies for the private anatomies (*Risposta dei Rettori di Padova ai Riformatori dello Studio intorno alle anatomie private*, in AS Ven., *Lettere degli Ecc. Riformatori dello Studio di Padova a diversi Ill. Rettori* [1601–1622], July 29, 1601, now in G. Sterzi, "Giulio Casserio," pp. 82–84).

7. "MDLXXXV, VII id. sextil. Jacobus Savoianus . . . , Prorector. Ex Litter. Ducal. Anatomica schola ad primariae fastigium evecta cum esset, reliquae vetitae sunt Anatomicae exercitationes, quae in privatis aedibus per urbem habebantur: qua de re saepe questa est scholarium Universitas per nuncios" (J. Facciolati, *Fasti*, pt. 3, p. 217).

8. *Stat. Gymn. Pat.*, l. II, chapter 28, pp. 90 ff.

treme penalties to be suffered as a consequence of it by all the students in the *Studium.*"⁹ In July 1601 the government authorities (*rettori*) in Padua replied to requests from university officials (*riformatori*) who had conveyed the displeasure of the entire faculty of arts, including medicine. The *rettori* communicated that they alone could grant permission for "particular" anatomies at the conclusion of the public sessions. This was because "if in general this freedom was to be granted to everyone, we would be blocking the public and regular one, causing it to terminate prematurely because of a shortage of bodies."¹⁰ These words seem to imply that the private anatomies performed until that time were on the fringes of legality, though they were tolerated by the institutions, enjoyed official status in Padua, and were recognized for their didactic function. Simultaneously, the rectors' reply gave physicians and students the opportunity to overcome the greatest obstacle to these supplementary lectures: the acquisition of cadavers.

If even before Vesalius private anatomies provided the students with an illegal but widespread and tolerated occasion to observe the human body closely (thereby making up for the deficiencies of the public demonstrations), anatomists, for their part, had other opportunities to examine the human structure and its operation. Medical writings, from the end of the fifteenth century, frequently alluded to numerous observations on the part of their authors at demonstrations at which they were present that occurred outside the usual public dissections permitted by the university statutes. The *De abditis nonnullis ac mirandis morborum et sanationum causis* by Antonio Benivieni (1443–1502), for example, is a pathology treatise that, in the course of a long survey of case histories, places a great deal of emphasis on the autopsy as a means of establishing the causes of death in patients or to study the anatomical and physiological changes brought on by disease. Benivieni reported that he himself had opened corpses, with the permission of their relatives, and testified to the widespread recourse to post-mortem examinations in Florence during the fif-

9. *Istanza dell'Università artista per ottenere le anatomie private,* in Biblioteca Universitaria, Padua, *Archivio Antico dell'Università,* vol. 677, fol. 208v, published by G. Sterzi, "Giulio Casserio," pp. 73–74. Another petition addressed to the doge (with a copy to the *riformatori* of the *Studium*), dated July 1601, is more precise in its requests and motives: " . . . la nostra Università si ha preso parte di supplicare, come fa, humilmente Vostra Serenità degnarsi di concedere, che fornito il tempo delle vacanze, deputato dalli Statuti nostri alla publica Anatomia, li scolari possino haver Anatomie private nelle case particolari di qualunque Dottore, per venir esercitando quello, che habbino imparato nella publica del sig.r Acquapendente, et in piú breve tempo imparar de quest'arte quello ch'è necessario loro per la medicina" (*Supplica per poter fare Anatomie private,* July 18, 1601, AS Ven., *Lettere*), and in G. Sterzi, "Giulio Casserio," p. 81.

10. AS Ven., *Lettere,* July 29, 1601, and G. Sterzi, "Giulio Casserio," pp. 82–84.

teenth century.[11] But even treatises dealing specifically with anatomy mention cases of autopsies that deviated from the etiological and pathological to become anatomico-physiological research. In his *Anatomice* the physician Alessandro Benedetti wrote that in Venice he had carried out research on the cadavers of syphilitic patients in addition to the cadavers he had cut open during the annual anatomies in the hospital of Saints Peter and Paul.[12] In the *Liber introductorius anatomiae* (1st ed. 1536), another Venetian physician, Niccolò Massa, mentions that he had opened a dozen or so cadavers. Two of these had been dissected in the hospital of Saints Peter and Paul, and so they must have been public anatomies.[13] The other bodies were probably not examined by Massa on those occasions fixed by the academic calendar. Three additional cases are of special interest because they provide an insight into the range of opportunities offered to the anatomist for observing the interior of the human body: the first concerns a Florentine nobleman, whose son, Bartolomeo de' Panzati, asked Massa to verify the cause of death; the second involves a "monstrosity against nature" born at Fossa Clodia; the third is that of a man given shelter in 1533 in the "convent of Saints John and Paul" at Venice for a head wound, whose recovery was deemed hopeless by surgeons and who was later dissected with the permission of the competent authorities.[14]

11. The book was published posthumously in Florence in 1507. Autopsies are discussed, for example, in chapter 3, pp. 33–37, 61, 76, 79, 81, 83, and so forth. Other evidence of this practice in Florence is provided by Leonardo da Vinci: "E questo vecchio, di poche ore innanzi la sua morte, mi disse lui passare cento anni e che non si sentiva alcun mancamento nella persona, altro che debolezza; e così standosi a sedere sopra un letto nello spedale di Santa Maria Nova di Firenze, senza altro movimento o segno d'alcuno accidente, passò di questa vita. E io ne feci notomia, per vedere la causa di sí dolce morte" (Windsor Castle, Royal Library, MS. 19027v). Théophile Bonet, in his *Sepulchretum Anatomicum* published in 1678, counted over three thousand cases of autopsies in the medical literature appearing between the end of the fifteenth and the first half of the seventeenth century.

12. The text was written in the 1490s. In the medical literature one frequently runs into presumed fifteenth-century editions of this writing, but I have never found one recorded in the incunable catalogues known to me. See also L. R. Lind, *Studies in Pre-Vesalian Anatomy* (Philadelphia, 1975), pp. 76–77. A doctoral thesis on Benedetti was defended by Giovanna Ferrari in 1991 at the University of Venice. For the dissection of a syphilitic woman, see Bk. I, chapter 6; for the annual anatomies, see Bk. III, preface and p. 36; Bk. IV, pp. 49–50, Bk. V, p. 68.

13. Massa, *Liber,* fols. 10r, 26r.

14. Ibid, fol. 56v: "Rogavi chirurgum ut post mortem vellet mihi denunciare, ut habita licentia a presidentibus ipsum anatomizarem, quod factum fuit, quoniam peregrinus erat, in qua dissectione plura vidi digna admiratione." On the noble Florentine, see chapter 27; for the monstrosity, chapter 28; for the case of another body dissected at S. Giovanni e Paolo, see chapter 14, and for the case of the two women, one of whom was eight months pregnant, see chapter 23.

In his *Commentaria* Berengario da Carpi frequently mentions his own examinations in Bologna on the cadavers of condemned men to corroborate the descriptions given in Mondino's *Anatomia* and in other texts in his possession. On March 17, 1520, Berengario in his home dissected the body of a woman in her ninth month of pregnancy.[15] It is not known how he obtained this cadaver, but there is no doubt that this was not a public anatomy and that the corpse had not been furnished either by the university or by the judicial authorities (a pregnant woman could never be executed).

It appears therefore that, even before Vesalius, the practice of dissection was much more widespread than the regulations contained in the university statutes would indicate. He stated that if one had to have young and robust bodies, resembling as closely as possible the ideal of the canon of Polykleitos for the public anatomies, for the other exercises, including those private anatomies performed with frequency (*"crebrius"*), no corpse should be refused: "it will be useful to tackle whatever [you get], that you may judge of what kind it is, and that you may understand the difference of bodies and the true nature of many diseases."[16] In the *Fabrica*, in fact, there are references to bodies of mysterious provenance that were used by Vesalius and his disciples for anatomical research, and to quantities of cadavers for public anatomies well in excess of what was permitted by the statutes. In addition to the anecdote about a skeleton stolen at Louvain, to which I shall return below, Vesalius also recalls the thefts of bones by young anatomists in hospitals and cemeteries, which even included the desecration of tombs.[17] Realdo Colombo, for his part, as evidence of the presence of a certain cranial suture, stated, with some exaggeration, that he had examined at least six hundred thousand skulls: some in Florence at the hospital of Santa Maria Nuova "where for countless centuries the bones of the deceased are preserved distributed in elegant piles," and some in Rome "in all the cemeteries for common people which are called *Campum Sanctum*."[18] In book 15 of the *De re anatomica*, entitled *De iis quae raro in anatome reperiuntur*, Colombo also mentions among his trophies autopsies performed on Cardinals Gambara (1540), Cibo (1550) and Campeggi (1554), and on the body of Ignatius of Loyola who died

15. I. Berengario da Carpi, *Commentaria*, fol. 260r.

16. A. Vesalius, *De humani corporis fabrica*, p. 548.

17. Ibid, p. 27: "Quando enim permulti effractis monumentis, atque ex xenodochiis conquisitis corporibus ossa ad ispectionem sibi parabant, facile accidit ut alius in senis, alius in pueri, ac rursus alius in huius illiusque aetatis hominum ossa inciderent, atque in scholas deportarent."

18. R. Colombo, *De re anatomica*, p. 19. Vesalius, too, relates that he had seen many lower jawbones in the foundlings' cemetery in Paris (*De humani corporis fabrica*, p. 44).

in Rome in 1556. A few lines before this account he tells of curious examinations carried out on monstrous bodies: two hermaphrodites, two Siamese twins, a certain Lazarus called "the voracious," and still others.

A stream of necrophilic boasting about the great number of bodies dissected and examined is found in anatomical books published between the end of the fifteenth and the mid-sixteenth century. It is not only the bodies of executed persons handed over by the judicial and university authorities that pass onto the dissecting tables, but also those of the sick abandoned in hospitals, of "monstrosities against nature," of those whose causes of death are not known or are suspicious. All this jars the limitations and the orderliness of the statutory norms imposed by the universities seeking to control the scalpels of the anatomists and the movement of cadavers. In contrast to the public demonstrations that had become solemn ceremonies emptied of any research or pedagogical potential, these private sessions, long before the time of Vesalius, presented quite other opportunities for physicians and students to begin a real process of familiarization with the handling of cadavers.

Authority and Evidence

" . . . those guided by the love of the truth, should gradually ripen therefrom, and give more credence to their eyes and not inefficacious reasoning, than to the writings of Galen."

A. Vesalius, *Fabrica*

In the history of anatomy from Mondino to Vesalius it is surprising that, despite claims of direct observation on cadavers that increased exponentially, the anatomical texts issued during this period do not add significantly to knowledge about the body and to the anatomical paradigm established by Galen and his followers who had lived in the West and in the Middle East for the past millennium. For us, intuitively, it would seem enough to look inside a body to distinguish its parts and then to describe them in some sort of writing. But it appears that it took at least two centuries, from 1315 to 1543, for this sequence of events to come about despite the fact that it was patently obvious. Why? First, because many authors rigorously followed the regulations established by the universities and limited the dissection to what was permitted in a public demonstration. For others, such as Benedetti, Achillini, Massa, Berengario, and Zerbi, who saw countless cadavers under the most disparate circumstances, the answer must be found in the fifteenth- and sixteenth-century texts themselves and, more precisely, in the attitudes of the writers toward the ancient authorities, whose overwhelming influence outweighed the evidence of the dissection. Here, too, the credit due Andreas Vesalius in the renewal that he brought to the study and teaching of this discipline will be apparent.

Teaching and writing on human anatomy was generally focused on Galenic medicine, on texts that were his or attributed to him, and on that mass of manuals, commentaries, treatises and encyclopedias inspired by them that translated, commented on, synthesized, or substituted for the original works. The *De anatomicis administrationibus* remained more or less neglected in Europe until the publication of the Latin translation by Demetrius Chalcondyles in 1529, although, as Berengario mentioned, its existence was known.[19] The latter, one of our most reliable sources on the bibliography of anatomy in the first decades of the sixteenth century, maintains that the only two texts of Galen's anatomy available at the time (1521) in Latin were an *Anatomia oculi* and an *Anatomia matricis*. Berengario had heard from "reputable sources" that Galen's as yet untranslated (into Latin) anatomical production totaled about forty titles.[20] Other writings by Galen, such as the *Methodus medendi, Tegni, De usu partium corporis humani,* which circulated widely in these years, even if they contained much anatomical information, were certainly not systematic anatomical treatises that could be used in teaching or learning the discipline, and even less as manuals of dissection.

In the anatomical field Galen dominated the textbooks, as he did the European universities, especially through the medieval writers who had ransacked his works. The third *fen* of Avicenna's *Canon* is an example of such borrowing, which would become a commonplace of university instruction. But the treatise that, more than any other, represented Galenic anatomy in the European universities for two centuries was undoubtedly Mondino's *Anatomia*. The history of anatomy to the time of Vesalius and beyond may be read as the history of the effort to escape the domination of this work.[21] Mondino's book, written in 1316, enjoyed rather quick success: as early as 1320 Stefano Arnaldi, vice chancellor of the medical faculty in Montpellier, was annotating and commenting on

19. "Et quamvis aliqui dicant Galenum Anatomiae ducem composuisse librum particularem in quo docet Anatomiam fieri adhuc de eo apud nos non est cognitio" (I. Berengario da Carpi, *Commentaria,* fol. IIIIr).
20. "Sed tantum duo reperiuntur libri apud nos de eius Anatomia particulari: primus est de anatomia oculi; alter de anatomia matricis. Verum ego quidem audivi a fide dignis quadraginta libros Galeni particulares reperiri nondum traductos in lingua latina de particulari anatomia a capite ad pedes quos non vidit Mundinus" (ibid, fol. IIIIr–v).
21. L. R. Lind, *Studies in Pre-Vesalian Anatomy,* p. 6.

it. The *Anatomia* became the teaching manual in that city in 1366 and just a few years later was adopted in Paris as well.[22] The Padua University statutes issued in 1465, which would be in effect for the entire sixteenth century and part of the next, called for the reading from Mondino's treatise as a guide during the public anatomies.[23] The same occurred in Bologna, as is evidenced by the anatomy course taught by Matteo Corti there in 1540.[24] As further proof of the important role played by this brief work, there are the many reprintings it enjoyed in the course of the fifteenth and sixteenth centuries throughout Europe, and the commentaries upon it prepared by noted anatomists of the time.[25]

Prevailing views, then, were profoundly colored by Galenism and Galenists, who were fiercely proud of methods and of knowledge that had survived for thirteen centuries. This enduring anatomical paradigm would eventually be displaced by dissection in a process that was to last more than half a century despite an intensification of anatomical activity and the prestige that the discipline began to enjoy from the close of the fifteenth century. Verification through a direct examination of the cadaver was alien to the objectives of the authors of most of the fifteenth- and sixteenth-century texts; the images of the anatomy lecture, represented by the quodlibetarian model, give iconographic form to this indifference.

22. J. Agrimi and C. Crisciani, *Edocere medicos. Medicina scolastica nei secoli XIII–XV* (Naples, 1988), p. 20.

23. *Stat. Gymn. Pat.*, l. II, chapter 28. Cf. J. Facciolati, *De Gymnasio*, p. 58.

24. "Porro Mundinum de more consueto in anathomia legemus" (B. Heseler, *Andreas Vesalius' First Public Anatomy at Bologna, 1540*, ed. R. Eriksson [Uppsala, 1959], fol. 22a).

25. The first printed edition of Mondino's manual was the *Anathomia* (Padua: Petrus Maufer [c. 1474]) (IGI, n. 5910). It was followed by Pavia: typ. Antonio Carcano, 1478 (IGI, n. 5911); Bologna: Giovanni de Nordlingen and Enrico de Harlem, 1482 (IGI, n. 5912); Padua: Matteo Cerdoni, 1494 (IGI, n. 5913); Padua: Antonio Carcano, 1492 (IGI, n. 5914); Venice: Bernardino Veneto, expensis B. H. Duranti, 1494 (IGI, n. 5915); revised "per Francischum Picium" Pavia: per Jacob de Paucisdrapis de Burgofranco, 1507; Lyons: in aedibus A. Blanchard, 1528; and Marburg: C. Egenolphus, 1541. There was an Italian translation: (Bologna, 1521), and a French translation: (Paris, 1532). See also *Anathomia Mundini emendata per doctorem melestat* (Leipzig [c. 1493]) (IGI, n. 5914). The *Anatomia* was also printed in the *Fasciculus medicinae*, edited by John of Ketham (Venice: G. and G. de Gregoriis, 1491) (other Latin editions 1493, 1495, 1513); Spanish translations: Pamplona: A. G. de Brocar, 1495, and Seville: J. Cronberger, 1517; Italian translation by Sebastiano Manilio Romano (Venice: G. de Gregoriis, 1508), and the one revised by Cesare Arrivabene (Venice: C. Arrivabene, 1522). For the commentaries see I. Berengario da Carpi, *Commentaria;* M. Corti, *In Mundini anatomen explicatio* (Pavia: F. Moschenus et G. B. Niger, 1550). One can also find a sort of commentary to Mondino in A. Achillini, *Expliciunt anatomicae [sic] annotationes Magni Alex. Achillini Bonon. Editae per eius fratrem Philoteum* (Bologna: Hieronymus de Benedictis, 1520). J. Dryander also prepared an edition, *Anatomia Mundini* (Marburg: C. Egenolphus, 1541).

The *Anathomia* of Girolamo Manfredi written in Bologna in 1490 opens with a greeting to the governor of the city, Giovanni Bentivoglio: "I have excerpted this opuscule as best I could from various ancient tomes, abbreviating them and not strictly following their order . . . "[26] The authorities in question were Avicenna and Galen, but also Mondino. The *Annotationes Anatomicae* by Alessandro Achillini published posthumously in 1520 were nothing other than a rearrangement of the lectures he had given in the first decade of that century on the basis of Mondino's *Anatomia*. Although the *Annotationes* were often accompanied by public demonstrations, the author in no way diverges from the text he is following except to indicate a few discoveries or, better still, rediscoveries. The arrangement worked out by Berengario in his *Commentaria* was much more complex. He juxtaposed to Mondino's text points of agreement or variance with what had been said by other writers, both ancient and modern.

The vast circulation of Galen's *De anatomicis administrationibus* made possible by the Latin editions of 1529 and 1531 brought about a redefinition of anatomy on the basis of the descriptions it contained. From these dates forward we find in every anatomy treatise an extraordinary (but not surprising) proliferation of citations from this work, to the detriment of other writers both ancient and modern, including Mondino. Galen's work, perhaps due to its length, was never accepted as a university text. Mondino's *Anatomia* remained, because of the way it had been put together, a more appropriate reading to accompany the anatomical demonstrations. In 1536, to make Galen's anatomy more accessible to his Parisian students, Winther published an abbreviated version of the *De anatomicis administrationibus* that he entitled *Institutionum Anatomicarum secundum Galeni sententiam*.[27] This work enjoyed great success until the early 1540s, as is demonstrated by the many reprintings in the principal European university and publishing centers.[28] In the introduction the teacher of Vesalius explained: "there is nothing written here that does not echo Galen, whose teachings I openly imitate: nothing is here asserted that I did not read there."[29] With these words Winther abdicated any will

26. G. Manfredi, *Anathomia* (1490), Bodleian Library, MS. 237, Western 20287.
27. See above, p. 29, note 36.
28. In 1536 editions appeared in both Basel and Paris. They were followed by Venice, 1538; revised by Vesalius, (Basel, 1539; reprinted 1541); and Lyons, 1541.
29. J. Winther (Iohannes Guinterius Andernacus), *Institutionum Anatomicarum secundum Galeni sententiam ad candidatos Medicinae libri quatuor* (Basel: B. Lasium and T. Platter, 1536), pp. 11–12. A similar statement is made by Andres de Laguna at the beginning of his brief compendium, *Anatomica methodus, seu de sectione corporis contemplatio* (Paris, 1535) (published in L. R. Lind, *Studies in Pre-Vesalian Anatomy*, pp. 263 ff.), where

on his part either to verify or revise any of the assumptions contained in the ancient text and discarded a priori the possibilities offered by other instruments of comparison, whether books he knew or the direct examination of cadavers he conducted.

In fact, the pre-Vesalian situation was not quite so desolate. The notion that Mondino, the Arabs, and even Galen might have missed something gradually made itself felt in the newer writings. Eventually more importance began to be afforded dissection. In the introduction to his *Annotationes,* Achillini defined anatomy as " . . . the skillful dissection and clarification of those things which lie hidden within the body." He was borrowing verbatim a passage from Johannes Alexandrinus's commentary to Galen's *De sectis ad eos qui introducendos.*[30] The same definition, explicitly attributed to Alexandrinus, is also used by Gabriele Zerbi in his *Liber anathomiae,* by Jacopo Berengario in the *Commentaria,* and by Niccolò Massa in the *Liber introductorius.*[31] A few years before, Alessandro Benedetti, in his dedication of the *Anatomice* to Maximilian I, affirmed that the entire theory and practice of medicine and surgery depended on anatomy. He then went on to say that no disease or alteration could be understood without a prior knowledge of morphology acquired through the dissection of the body.[32]

Even though none of these writings attempted any serious criticism of the ancient authorities and of Galen in particular, definitions of this type, where the emphasis was placed on the potential of the manual aspects of dissection, and its actual identification with anatomy that transpired from these statements, anticipate the change. This becomes most explicit in Berengario's *Commentaria,* which corrects Galen's anatomy in various

---

he says: "I shall not say anything in this commentary that cannot be proved by the authority of Hippocrates, Galen, Cornelius [Celsus], Plato, Aristotle, Pliny or, finally, Alexander of Aphrodisias. Beware, however, because if my discourse fails to persuade you, the credibility and authority of these illustrious men is inevitably thrown into doubt" (p. 265).

30. The passage by Iohannes Alexandrinus reads: "Anathomia est artificiosa incisio et clarificatio eorum quae in occulto ascondita sunt in corpore. Artificiosa incisio dicta est eo quod medici docte et articulate indicant quae foris sunt artarias [sic]. Clarificatio quae in occulto sunt bene dictum est quia in hoc non possumus scire nisi per anathomiam." This is also contained in a text that introduces the Latin translation of Rusticus Placentinus's edition of Galen's *Opera omnia* (Pavia: J. de Burgofranco, 1515–16), fols. 6r–12v. On Alexandrinus's commentary, see O. Temkin, "Studies on Late Alexandrian Medicine: I. Alexandrinus' Commentaries on Galen's *De sectis ad Introducendos,*" *Bulletin of the Institute of the History of Medicine* 3 (1935): 405–30, and R. French, "A Note on the Anatomical Accessus of the Middle Ages," *Medical History* 23 (1979): 426–63.

31. G. Zerbi, *Liber Anathomiae corporis humani et singulorum membrorum* (Venice: B. Locatellum, 1502), preface; Berengario da Carpi, *Commentaria,* fol. Vir; Massa, *Liber,* fol. 3v.

32. A. Benedetti, *Historia,* letter to Maximilian.

ways. The entire introduction to the work is a display of anatomical methodology and epistemology that expounds on the criteria adopted by Berengario to research, comment on, and teach the discipline.[33] For the first time there is a questioning attitude toward textual authority that is clearly expressed and is dictated by a view of science that would become dear to Newton: "We know how to carry out science by adding parts to other parts: and we are as children standing on the shoulders of giants: we are able to see much farther, than antiquity could."[34] With this aim in mind the good anatomist should first of all be "educated in the science of medicine" and have read what Aristotle, Galen, Avicenna and others had written about animals living and dead. However, facts so acquired should not be slavishly accepted: "And in this discipline nothing is to be believed that is acquired either through the spoken voice or through writing: since what is required is *seeing* and *touching*." In fact, Berengario goes on to stress that Galen and the other writers are not to be followed "where experience and judgment are in opposition." Nor are one's own observations to be trusted "but if possible the authority of the doctors and one's own views should be tested through experts in anatomy."[35] The *Commentaria*, as I have mentioned, repeat in plan the usage adapted for this literary genre: Mondino's text is divided into chapters, each of which is followed by a long commentary by Berengario in which he shows where the work corresponds to or contradicts other authorities, both older and more recent, as well as his own observations made during a dissection. In several places, on the basis of these findings and of reason (the logic to which Galen had referred), he corrects Galenic authorities, placing the most confidence in knowledge acquired through sensory perception.[36]

Similarly, Niccolò Massa, author of the *Liber introductorius anatomicae* published in Venice in 1536, took his cue from Mondino and followed the *Anatomia*'s organization in describing the parts of the body, even though he had available and quotes frequently from Galen's *De ana-*

33. The author asserts, principally, the utility and characteristics of anatomy—contemporaneously *instrumentum medicinae, alphabetum medicorum, ars* and *scientia*—and its simultaneous association to the fields of practical and theoretical medicine, as well as to philosophy (I. Berengario da Carpi, *Commentaria*, fol. Vr–v). For the physician, he specifies, "anatomia est necessaria non solum in cognoscendis aegritudinibus verum etiam in curandis et etiam in praeservandis et conservandis corporibus in sanitate."

34. Ibid, fol. IIIIv. This metaphor is attributed to Bernard de Chartres by John of Salisbury, *Metalogicus III*, 4, ed. C. J. Webb (Oxford, 1929), p. 136.

35. I. Berengario da Carpi, *Commentaria*, fols. VIv and VIIr.

36. Ibid, fol. CCCCXIIv: " . . . Galen cum suis sequacibus cuius opinionem semper tenemus nisi ubi discordat ab ipso sensus." In another already cited passage, he directs himself to the authors of anatomical treatises: " . . . ergo sint cauti componentes libros de anatomia et non credant auctoritatibus: sed sensui sicut nos facimus et faciemus" (fol. CLIIIv).

tomicis administrationibus.[37] Unlike Berengario he does not give the names of more recent writers on anatomy, against whom he hurls invectives: "They are read and considered oracles, they who either have handed down . . . a mutilated knowledge of this enterprise [anatomy], or have described the writings of others, which they understood badly, and have decorated them with sophistical arguments. Dissolving with glosses the questions at stake between Aristotle and Galen, they have obfuscated the mind of the young with distinctions to such an extent that you will find almost no one who understands (*sensatam*) anatomy completely." And a few lines later, he continues with the same vehemence: "I am astonished to say (God is my witness) how great is the ignorance of some moderns not only in their speech but also in their sensations; they are always paying lip service with their mouths to Aristotle, Hippocrates, Galen and Avicenna, although they in fact know nothing certain, but like little birds in speaking and chattering what they hear, delight the ears of a few." Not only that, but they had dared to write even about those things "that they had not seen with their own eyes, nor touched with their hands."[38] In contrast to this backward attitude toward learning still endemic in academic circles, Massa insisted, as Berengario had also done, on the use of sight and touch, in other words on sensory experience. Like Berengario, in fact, Massa considered anatomical science a progressive accumulation of knowledge. This did not mean that he felt free to attack or slander the ancient authorities, but instead to continue along the path they had traced:[39] the most appropriate way to learn about and to teach the struc-

37. " . . . quanquam . . . multi sint modi aggressionum, modum, quem Mundinus vir in sectione celeberrimus tradidit, insequar" (Massa, *Liber,* fol. 5r).

38. Ibid, fols. 3v–4r. This first chapter is addressed to "Gerolamo Marcello," probably to be identified with Gerolamo Ferro. Cf. L. R. Lind, *Studies in Pre-Vesalian Anatomy,* p. 174, note 2. It is interesting to note the close resemblance between Massa's words and the preface to the *Fabrica* of Vesalius (1543) where the contemporary teachers of anatomy are also compared to birds, crows to be precise, who repeat from memory what they have learned from the books of others. Similar notions are expressed by Valverde in his "Letter to the Readers" in which he compares professors to trumpets: "Sí che io consiglierei ciascuno o a venire qui in Italia dove la potrà facilmente vedere [dissection], o se pur ha da stare al detto di coloro che ne scrivono, che voglia piú tosto credere a coloro che hanno speso tutta la loro vita in questo studio con maggior copia d'huomini, che non poté mai Galeno haver di simie, che a coloro, che come trombette dicono dalle cattedre, non quello ch'essi hanno veduto, anzi quello, che pur ogni picciolo fanciullo potrà per se stesso leggere havendo il libro innanzi" (*La anatomia,* preface).

39. "Nolo te tamen credere me antiquos: a quibus edoctus sum, calumnis prosequi: immo ipsos semper laudo: quoniam sunt laude digni etiam qui imperfecta principia aut partes doctrinae tradiderunt, cum nobis postea addendi aliquid ansae dederint." A similar statement is repeated later: "[G]rande etenim malum est, doctissimos, a quibus tot bona scripta sunt, lacerare: et si aliqua etiam dixerint, vel obscura, vel non ab omnibus approbata. Equidem Platonem laudo: Aristotelem, Hippocratem, Galenum, Avicennam, Averroim et alios doctissimos, qui posteritati tantopere suis laboribus profuere" (Massa, *Liber,* fol. 4v).

ture of the body was by utilizing every available method: "the reading and difficult exercise (of dissecting the body) of the ancients."[40] Hippocrates, Galen, and Aristotle were also mere men and as such liable to make mistakes or to forget something,[41] so it was better to choose, within the limits of respect owed to the sages of old, "the truth over the authority of men." Truth (*veritas*), in this case, was unequivocally defined as that which is perceived through the senses, that is discovered by the anatomist through the examination of the dissected. Massa concludes his introduction with the advice: "so that you may learn everything, you should often practice dissection, in the way I shall explain below, or in some other if you find a better way."[42]

LIMITATIONS OF BELIEF: VESALIUS, GALEN, THE GALENISTS

Niccolò Massa's *Liber introductorius anatomicae*, in spite of its invectives against ignorance and conformity and in spite of the propositions set out in the introduction, contains little that is new with respect to his classical or more recent predecessors such as Berengario's *Commentaria* and the works of others writing before Vesalius. The *panniculus carnosus* that he claims to have discovered had already been detected by Avicenna, and even the two muscles in the lower part of the throat had already been isolated and described.[43] Massa also went to great pains to provide a minute description of the *rete mirabile*, although he admitted that it was often difficult to discern it in men since, being very fragile, it disappeared quickly after death.

It is necessary to comment briefly on this *rete mirabile*: the descriptions of it furnished by Renaissance anatomists eloquently exemplifies their attitude toward the ancient authorities.[44] In Galen's anatomic-physiology, as in that of Herophilus before him, the internal carotid artery separated itself at the base of the skull into a thick mesh of tiny blood vessels. Here the vital spirit (*spiritus vitalis*) contained in the blood flowing from the

---

40. This suggestion of method had already been made by Galen in *De anat.*, K II 220 (see above, p. 139). Guy de Chauliac, from his reading of the *De anatomicis administrationibus*, echoes the words of the ancient authority in urging a study of anatomy "per librorum doctrinam" and "per corporum mortuorum experientiam" (Guy de Chauliac, *Cyrurgia,* fol. 5v).

41. "Sed non sum ille: qui dicam homines sine errore naturam genuisse, quare, cum homines fuerint, potuerunt et ipsi omnes errasse" (Massa, *Liber,* fols. 4r–v). Then he resumes: "ipsi aliquando erraverunt, neque potuerunt omnia cognoscere" (fol. 5r).

42. Ibid, fol. 7r.

43. L. R. Lind, *Studies in Pre-Vesalian Anatomy,* p. 11.

44. On the *rete mirabile,* see C. Singer and R. Rabin, *A Prelude to Modern Science: Being a Discussion of the History, Sources and Circumstances of the "Tabulae anatomicae sex" of Vesalius* (Cambridge, 1946), pp. xliii–xlv.

heart became transformed into animal spirit (*spiritus animalis*), and it circulated in the body through the nerves. The *rete mirabile* thus provided the link between mental processes and muscular movement, between soul and body; it was the key to understanding one of the most mysterious and intriguing aspects of nature. In fact a similar network of vessels is discernible in hoofed animals, though its function is obscure, and, to a lesser degree, in carnivores, but there is no trace of it in humans, monkeys, or rodents. Thus it is easy to see why Massa had trouble locating it!

Although Berengario wrote that he had never seen the *rete* and that, contrary to what was maintained by other contemporary anatomists, he thought it did not exist, he offered a detailed description of it.[45] The *rete* is also portrayed in the third of Vesalius's *Tabulae anatomicae* published in 1538 (fig. 34).[46] A few years later, however, he realized that he had described something that did not exist in humans. In the *Fabrica* he explains that the error was due to the method he had used in his anatomical demonstrations: he had always dissected human heads together with those of lambs and cattle so as to be able to show in the latter what could not be seen distinctly in the former. Consequently, trusting in the veracity of Galen's *De usu partium* and in the comparative method he had borrowed from him, he had foolishly erred.[47]

Sixteenth-century anatomists had therefore not strayed much beyond the ancient masters to the extent that, at times, the force of tradition compelled them to reject the available evidence: they often did not believe what was actually before them. But they took a big step forward, nevertheless, when they recognized, at least in principle, the priority of observation over authority, even if the consequences of this did not immediately become apparent. Vesalius's activity and his writing need to be viewed within this intellectual and material context, which made the *Fabrica* and the ensuing renewal of anatomy feasible. Looking at him and his work in this way should not detract from his merits, however. Vesalius went beyond assertions of principle and directed his criticisms not only against contemporary anatomists and academic custom (the decline of anatomy, the empty anatomical ritual, the disdain for manual operations), but also expressly against the main source of anatomical knowledge: Galen. In the

45. I. Berengario da Carpi, *Commentaria,* fol. CCCXLVIIv. Cf. G. Rath, "Pre-Vesalian Anatomy in the Light of Modern Research," *Bulletin of the History of Medicine* 35 (1961): 142–48.

46. The *Tabulae,* printed in Venice in 1538, are now also available in C. Singer and R. Rabin, *A Prelude to Modern Science.* At the letter "B" of the third plate we read: "Plexus reticularis ad cerebri basim, Rete mirabile, in quo vitalis spiritus ad animalem praeparatur."

47. A. Vesalius, *De humani corporis fabrica,* p. 642.

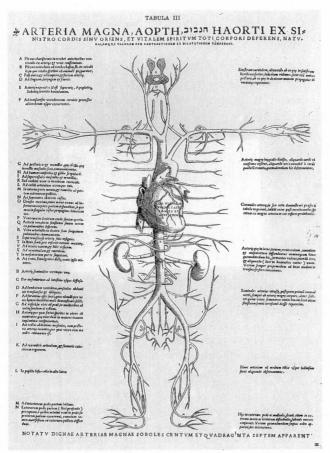

Figure 34   Andreas Vesalius, table III, "Arteria Magna," from *Tabulae Anatomicae Sex* (Venice, 1538). In 1538 Vesalius published six anatomical tables in Venice: three of these, executed by Johann Stephen van Calcar, portray the skeleton; the other three, drawn by Vesalius himself, are diagrams of human anatomo-physiology depicting the liver and the portal vein, the circuit of the vena cava, the heart, and the aorta. In this latter figure Vesalius pictures the *rete mirabile*, a network of minute blood vessels that, according to the dictates of Galenic theory, transforms the *spiritus animalis* into a *spiritus vitalis*. No such reticulum exists in the human body. In the *Fabrica* Vesalius acknowledges the error and his ingenuous acceptance of Galen's description.

earlier section devoted to the title page of the *Fabrica,* I showed how he labored over the methodological and content deficiencies of the Galenic paradigm in the preface to that work. In the index alone, incomplete as it is, Vesalius referred to Galen 265 times in the course of describing the parts of the human body, in the majority of instances to correct him, particularly as his examinations had been carried out on animals and not on humans.

To understand Vesalius's position regarding the relationship between authority and evidence, instead of dwelling on the corrections to Galen's anatomy contained in the *Fabrica* it is useful to note significant passages in the controversy that arose between Vesalius and Matteo Corti in the course of anatomy lectures that they held in Bologna in the winter of 1540. The episode is recounted by Baldasar Heseler. The trouble erupted over a difference of opinion between Galen and Hippocrates concerning the best way to cure certain pleurisies on the basis of anatomical considerations. The disagreement had been noted by Corti in his morning lecture. The scene took place during the afternoon anatomical demonstration held in the presence of over five hundred students:

> When Corti's lecture was over Vesalius came and heard the confutation of his arguments, and he asked Corti to accompany him to the anatomy. Vesalius wanted to show Corti that his own opinion was most true. He therefore led Corti to our two subjects: "Behold, most excellent sir," he said, "we have here two subjects. We shall now see whether I have erred. We want to look at them, and to leave Galen aside for the time being, for I confess that I have stated (if one may say so) that Galen has erred here, since he did not know the location of this vein, without *coniugatio* in the human body, which we have today just as he had it." Corti answered, smiling, for Vesalius, being angry, was very agitated: "We must not," he said, "abandon Galen, since he always understood everything well. We therefore also follow him here. Do you know how to interpret Hippocrates better than Galen did?" Vesalius answered: "I did not say that, but here I show you, in these subjects, this vein without a counterpart [this unmatched vein], how it feeds all the inferior ribs, except the two upper ones, in which there is no pleurisy." Corti answered: "I am no anatomist, but there could well be, in addition to these, still other veins nourishing the ribs and muscles." "Where, I ask you?" said Vesalius. "Show them to me." Corti said: "Do you want to deny that there are natural passages?" "Oh," said Vesalius, "You want to speak of nonappearances and the occult; we, however, want to speak of manifest things." Corti answered "I always deal with the most manifest things; you, sir, do not understand Hippocrates and Galen [when they treat] them." Vesalius replied: "Very true, for I am not an old man like you, etc." They assailed each other in this manner with much quarreling and quibbling, and all the while got nowhere. Vesalius said: "Doctor, your excellency should not think me so dimwitted that I do not know or understand these things." Corti replied, smiling: "Sir, I did not say this, for I have always called you excellent, but I have reproved the false exposition of Hippocrates, that Galen erred in

these matters." Vesalius answered: "I confess that I did claim that Galen erred in these matters, and this is obvious here, in these subjects, as are many other errors of his as well."[48]

The dialogue between the two anatomists seemed to articulate with particular immediacy both the position of Vesalius, who was ready to question Galen's authority on the basis of the evidence, and the conservative posture of Matteo Corti. Corti told the students during the lecture inaugurating his Bolognese course that anatomy, both etymologically and in reality, meant dissection or the division of the parts of the body into several sections or even "repeated incision." The *sectio* (dissection), Corti added, can take place in two ways: "one in fact and immediate, the other through discourse or writing and speaking about it; for this also is to dismember the body." This art, the lecturer continues, was highly imperfect in Aristotle's day, but became "highly perfected and complete" with Galen, as is demonstrated by his anatomical writings, especially the *De anatomicis administrationibus*. Corti, along with most of his contemporaries, began from the presupposition that Galen had long practiced dissection even on the human body and that what he had written was incontestable because it had been founded on observation and reason. This attitude did not exclude the possibility of making new discoveries through dissection. Mondino, for example, according to Corti, "added much through anatomy," but then qualified, "all of which, as you shall see, was false." Corti's blind faith in Galen, finally, was revealed in the regret with which he noted the loss of five or six books of the *De anatomicis administrationibus*. They "have rendered anatomy much more perfect in our own times, than it was at the time of Galen; yet, however perfect [the anatomy] we have, it is not as perfect as it was at the time of Galen."[49]

In contrast to this fideistic and indolent acceptance of Galenic dogma stands the demonstrative obstinacy of Vesalius who, at various points in the dissections he was holding at that time in Bologna, found himself disagreeing with Galen, and therefore, inevitably, with Corti. The *Fabrica*, published with the *Epitome* three years later, in 1543, was intended to correct the errors that had been passed down for a millennium. Vesalius did not, however, ignore the decisive role Galen had played in the history of anatomy and medicine and even defined him in his preface as "the prince of medicine after Hippocrates." Vesalius recognized that it was

48. B. Heseler, *Andreas Vesalius*, p. 272.
49. Ibid, p. 46.

only after he had steeped himself thoroughly in the Galenic texts that he was in a position to write his own *magnum opus*. It is not unusual to run into errors that even he had appropriated from the ancient tradition: a notorious example of this is Vesalius's acceptance, in the absence of other explanations, of Galen's description of the heart, according to which the flow of the blood from the right to the left ventricle takes place through pores visible in the central septum.[50] Another tribute to Galen was the choice of anatomical terminology and the order followed in the exposition of the parts of the human body, all borrowed from the *De anatomicis administrationibus* (osteology, myology, veins and arteries, nerves, and organs), rather than from Mondino's organizational plan (lower, median and upper abdomen, from the exterior to the interior) normally used in fifteenth- and sixteenth-century anatomical writings.[51]

Implicitly, Vesalius's critique of Galen was also an attack on all those who had blindly borrowed from him and been inspired by him. Only rarely does the *Fabrica* mention more recent authors. Even Mondino is only cited once.[52] There are frequent allusions, however, to the names and works of the most ancient writers, of those who, for better or worse, represented authority: Galen, naturally, is in first place, followed by Hippocrates, Aristotle, Plato, Celsus, Pliny, Oribasius, and Theophilus, who are joined by the Arab scientists Avicenna, Razi, Mesuè, and Alí Abbas. Finally Vesalius mentions all those anatomists who are known only through the writings of others and who, for him, represent the golden age of anatomy: Herophilus, Erasistratus, Diocles, Marinus, and Lycus. With the obvious exception of this last category, all, and especially contemporaries (whose names are prudently suppressed), had failed to subject Galen's writings to verification. This had occurred for two reasons: because of the attitude of physicians toward Galen's works (which they all trusted), which made it impossible for them to believe he could ever have erred, and because "never have they seen anything as offensive as the dissection of human bodies."[53] In fact this is the point that set Vesalius apart from Corti and the majority of contemporary anatomists. Through dissection and a careful reading of Galen's works Vesalius had concluded that the latter had never practiced dissection on the body and had based his entire anatomy on the observation of monkeys. Consequently Galen's

50. A. Vesalius, *De humani corporis fabrica*, p. 589.
51. "Quippe in horum librorum ordine digerendo, Galeni sententiam sum sequutus, qui post musculorum historiam, venarum, arteriarum, nervorum, et dein viscerum Anatomen pertractandam duxit" (ibid, fols. 3v–4r).
52. Ibid, p. 531, apropos the formation of the fetus.
53. Ibid, fols. 3r–v.

greatest error had been in describing human anatomy without considering the multiple differences that distinguished the human body from that
of the primates.[54]

These words and the entire *Fabrica* unleashed a bitter conflict with the
defenders of the Galenic tradition. Vesalius himself, being well aware of
the innovations contained in his treatise and of the academic and intellectual conservatism that surrounded him, had predicted it.[55] Jacques Du
Bois (Iacobus Sylvius), who, as we recall, had been his teacher, was the
first to align himself with the traditionalists in a bitter and violent pamphlet entitled *Vaesani cuiusdam calumniarum in Hippocratis Galenique
rem anatomicam depulsio* (Paris, 1552). This was followed by Francesco
Pozzi's *Apologia in anatome pro Galeno, contra Andream Vesalium
Bruxellensem* (Venice: F. de Portonariis, 1562) and the *Apologia Francisci
Putei pro Galeno in Anatome examen* by the Milanese physician Gabriele
Cuneo (Venice: F. de Franceschi, 1564). On the opposite side of the bar-
ricades, alongside the anatomical writings of many contemporaries who
recognized and defended the role of Vesalius and the value of the *Fabrica*
(for example, Colombo, Valverde, Falloppio, Fuchs, and Coïter), stood
Renatus Henerus, author of an apology supporting the Flemish anatomist
in which he responded to the *depulsiones* of Du Bois.[56]

Among the physicians, Du Bois was probably the most significant expo-
nent of the conservative academic tradition. His teaching in the University
of Paris medical faculty was obsequiously modeled on the medicine and
anatomy of Galen and Hippocrates of whom, it could be said, he was
the official spokesman in France. In fact, on the title page of his *Vaesani*
he called himself "Royal interpreter of things medical among the Pari-
sians."[57] Vesalius's *Fabrica,* in addition to casting doubt on Galen's anat-

54. "Quinetiam quamplurima apud Galenum invenias, quae in simiis quoque minus recte assecutus est, ut taceam, mirandum esse maxime, in multiplici infinitaque humani corporis organorum et simiae differentia, nullam nisi in digitis ac poplitis flexu, Galenum animadvertisse: quam cum caeteris procul dubio omisisset, nisi citra hominis dissectionem ipsi
fuisset obvia" (ibid, fol. 3v).

55. "Licet interim non me lateat, quam conatus iste meae aetatis, qua vigesimum octa-
vum annum nondum excessi, nomine, parum autoritatis habebit, ac quamminime ob cre-
bram non verorum Galeni dogmatum indicationem, ab illorum morsibus erit tutus, qui mihi
Anatomen docenti non astitere, aut ipsi hanc sedulo non sunt aggressi, primaque fronte
varias rationes in Galeni defensionem comminiscentur, nisi magno alicuius numinis patroci-
nio ex more commendatus in lucem auspicato prodeat" (ibid, fol. 4r).

56. *Adversus Iacobi Sylvii depulsionum anatomicarum calumnias, pro Andrea Vesalio
Apologia* (Venice: [Gualtiero Scoto], 1555). Du Bois's text, from which the quotations that
follow are taken, is reprinted in the volume.

57. The same definition can be found in an anatomical compendium based on Hippocratic and especially Galenic writings that he published in Paris in 1542 with the title *In
Hippocratis et Galeni physiologiae partem anatomicam isagogae.* It would be reprinted in

omy and to proposing a comprehensive reexamination of related cognitive strategies, was thus undermining the academic prestige and credibility
of Du Bois and all those who based their teaching entirely on the ancient
tradition. Vesalius is never mentioned in the *depulsiones,* but in the title
itself the derisive epithet "*Vaesanus*" (mad, insane) is placed as an obvious allusion to "Vesalius."[58] Du Bois accuses him of slandering Hippocrates and Galen and of having polluted anatomical science, physiology, and
medicine in general by casting doubt on the truth of their writings.[59] Du
Bois took it on himself to respond both to the specific question of Galen's
errors pointed out in the *Fabrica,* and to Vesalius's basic thesis, namely,
that this ancient authority had never conducted dissections on the human
body, but only on animals and in particular monkeys. Fearing that Vesalius, whom he compared to the plague, would irrevocably contaminate
the thought, the activity, and the works of others (as had already happened),[60] Du Bois appealed first to Charles V, asking him to condemn and
put a halt to this "most pernicious example of impiety,"[61] and then to all
"the pious reborn French, German and Italian Aesculapians," urging
them to rise in defense of the hallowed masters.[62]

   Du Bois in his *Vaesani* does not hesitate to link, without stressing the
point, the use of dissection with his vigorous defense of textual authority.[63] In his *Isagogae,* a work written some years earlier and conceived as

---

Paris in Latin and in French by Jan Hulpeau in 1555; in 1556 in Basel by Derbilley; in
Venice by Valgrisi; and once again in Paris in 1560 and 1561 by the Gorbin printing establishment, together with a commentary to the *De ossibus* and a text in which he indicates
the order to be followed when reading the writings of the two ancient authorities (*Ordo et
ordinis ratio in legendis Hippocratis et Galeni libris*). I have used the Venice 1572 edition
of the *Isagogae* entitled *Institutiones anatomicae. In quibus ea omnia quae ad rem anatomicam bene docendam pertinere possunt, brevi ac facili modo ostenduntur.*
   58. R. Henerus (*Adversus,* p. 2) promptly spotted Du Bois's rhetorical artifice: "vero
nomen supprimit, et pro Vesalio Vaesanum substituit, singulari cuidam prudentiae."
   59. "Divinum Hippocratem, ac eius ubique admirandum interpretem Galenum, maxima
post Apollinem ac Aesculapium medicinae numina, cum mihi persuasissem undequaque
absolutissimos esse, nihilque unquam nec in physiologia, nec in caeteris medicinae partibus
scripsisse, quod non esset verissimum, audiissem vero vanissimas quasdam in ipsorum physiologiam calumnias a vaesano quodam ac arrogantissimo simul et rerum omnium igno-
rantissimo transfuga, iniquissime iactari, omnem movi lapidem, ut pium auditorem decebat,
quo eos omni ignorantiae non modo nota, sed etiam suspicione liberarem" (J. Du Bois,
Vaesani, p. 73).
   60. J. Du Bois uses the example of Fuchs, *De humani corporis fabrica ex Galeni et Andreae Vesali libris concinnatae, epitomes pars prima* (1551) (ibid., p. 74).
   61. Ibid., p. 133.
   62. Ibid., p. 134.
   63. Ibid, p. 133.

an anatomical teaching aid on the Galenic model,[64] the Parisian physician had stressed the didactic utility of this technique in a chapter entirely devoted to it:[65] "It is simple for the purpose of teaching and is briefer, surer, more certain and more conducive to remembering." Agreeing with what had already been said by many anatomists, he affirmed that " . . . dissection is the way for you to experience through sight and touch, rather than through hearing and reading." The assumption behind this statement is set out a few lines later, "since the book of nature immediately places everything before the eyes, which are more worthy of faith than the ears."[66] Du Bois, in fact, invited the reader to examine corpses personally, even "touching them with your hands," and devoted the last part of the book to the description of a series of public and private anatomies carried out on human bodies and on animals during his career.[67]

This may surprise us. How can we reconcile the vigorous defense of the ancient anatomical tradition with the practice of dissection, which is even outside the restrictions of academic ritual? How can Galen's errors be justified when the mere fact of dissection exposes them? More generally, how was it possible to still defend this tradition in the face of the evidence proffered by Vesalius, one so solidly based on the direct examination of the cadaver? Du Bois himself attempted a reply: along with every other physician of Galenic persuasion he was convinced that the author of the *De anatomicis administrationibus* had himself carried out numerous observations on human cadavers. It was inconceivable at mid-sixteenth century that any anatomist, Galen included, could have avoided this practice: at stake was the very credibility of his work. The entire battle waged by the Galenists against the anti-Galenists pivoted around this point.[68] Galen's descriptions, when tested by the anti-Galenists, had often turned out

64. J. Du Bois, *Isagogae*, fol. 11v: "Favete igitur animis et linguis lectores optimi, et diligenter attendite, dum vobis Galenicam puritatem quam possum brevissime et clarissime interpretor."

65. Ibid, fol. 93r. Chapter 23 on the subject of dissection is entitled "*De administratione Anat.[omica]*. Here the author indicates what the ideal morphological characteristics are for cadavers marked for dissection. He also emphasizes the importance of carrying out the practice as often as possible even on bodies that do not fit the ideal criteria and on the carcasses of animals, in addition to providing technical advice on the instruments to be used (fols. 92v–97r).

66. Ibid, fols. 94r, 96v.

67. Ibid, fols. 97v–106v. This part is entitled *In variis corporibus secandis observata quaedam a Iacobo Sylvio medico.*

68. In his pamphlet against Sylvius, Renatus Henerus wrote: "Galenus nusquam, quod sciam, humanae a se visae aut factae dissectionis, praeter iam dicto loco, meminit, silentio neutiquam praeteriturus, si qua alia unquam ipsi contigisset" (*Adversus*, p. 7).

to be wrong, and Du Bois himself, an active dissector, would have been aware of this fact. Consequently, to prevent the disintegration before his very eyes of the entire anatomical system based on the Galenic tradition, he had to explain the discrepancies between the descriptions of the cadavers recorded by Galen and those of sixteenth-century anatomists. Du Bois therefore maintained that human bodies, like those of animals, underwent changes in the course of the centuries; similarly, different geographical origins could lead to different morphologies: "Their writings and our bodies abundantly testify that in many regions the size, number, shape, and location of some of the internal organs have changed, nor do the same things that the ancients observed remain in existence in all our own bodies, so that it is believable either that something peculiar inheres in the men, as well as the animals and plants of specific regions, or that some change from an earlier nature has occurred."[69] Galen's total rehabilitation followed from this and provided one more reason to pursue research in this field with alacrity: "He recorded what the ancients and he himself saw very frequently or always: he never taught anything false or close to a lie, as all of his books most amply testify. Our elders [the ancients] are therefore not to be taxed with ignorance or neglect in anatomy. But the cause of change in our bodies, a cause that vitiates the nature of bodies, is not to be identified."[70]

This evolutionary view of human history *ante litteram* in an attempt to reconcile ancients and moderns over the issue of anatomy is distinctly farfetched. In the controversy between Galenists and anti-Galenists it was not only Galen's reputation that was at stake, but the credibility and prestige of the contenders, with all that could ensue from the point of view of their intellectual, academic, and social standing. The stakes were high. Thus it seems reasonable to think that even Du Bois did not give full credence to what he had written and that his crude defense of Galen was really a last-ditch effort against all the evidence. To verify such a supposition we would have to measure how much Du Bois and other anatomists before him *believed* in Galen and in the infallibility of the ancient authorities.

In fact it is precisely in terms of credence that Du Bois posed the problem of attitude toward the ancient texts. The failure to revise Galen's writings during the years in which dissection and the direct observation of cadavers had become ever more popular and efficaciously practiced is in fact comprehensible only if we consider the fideistic character of this ac-

69. J. Du Bois, *Isagogae*, fol. 6v.
70. Ibid, fol. 11r.

ceptance of the master. Du Bois, in the *Vaesani* as well as in the *Isagogae*, employs language more suitable for a religious than a scientific argument. The posture of the anatomist and of students in respect to the "divine" Hippocrates and the "pious" Galen should be dictated principally by "piety" and "faith."[71] Against them are arrayed the "impiety" of the slanderers who dare to contradict, with "sacrilegious mouth,"[72] Galen's anatomical texts designated as "sacred discourse."[73] Du Bois appeals to the reader (always "pious reader" in the *Vaesani*), the student, and the physician to stir them toward establishing such observance of the truth as is revealed by the "divine teachers" and opposing future slanders, in a tone suitable to the most ferocious Inquisitorial tribunal: "If this wild beast of Lerna [the Hydra] carries some new head, you should promptly cut it down, and you should mangle, trample on and offer to Vulcan this Chimera of a false conception, this crude and disorganized cesspool and sewer of hodgepodge, a work unworthy to be read by you, even if it is dressed in some feathers of Galen, but polluted by the filth of barbarism so that the splendor of the author has been debased."[74] Further, Du Bois did not hesitate to define as "false dogma" the opinions of those who had abandoned true doctrine. He hoped that physicians and students, with the aid of his own works that restored Galen's anatomy to their original "purity," would learn "not to despise their divine masters."[75] In fact, Du Bois wrote, concluding the preface to the *Isagogae*: "and I urge, as they come to an appreciation of Galen's anatomical truth through our labors, to return to believe in him as soon as possible."[76]

The fideistic attitude toward the ancient written authority displayed in Du Bois's language and rhetoric, and his appeal for a careful reading of the "sacred" texts of anatomy under the guidance of official interpreters for the purpose of reestablishing the truth, were certainly not unique to him. Even the author of the *Opuscula anatomica*, Bartolomeo Eustachio, attacked Vesalius in almost religious tones in an unpublished text entitled *Anatomica Tractatio ex Vesalio et aliis Anatomicis*. He maintained that it was not so much a question of Vesalius's bad faith in interpreting Galen,

---

71. R. French has also dealt with these themes in "Natural Philosophy and Anatomy," in *Le corps à la Renaissance. Actes du XXXe colloque de Tours, 1987* (Paris, 1990), especially pp. 450–54.

72. J. Du Bois, *Isagogae*, fol. 11v.

73. Idem, *Vaesani*, p. 74. To this Henerus responded: "Neque enim sacer ille sermo polluendus erat nefario nomine: et recte quidem, quandoquidem sacrum ille appellat, quod omnes alioqui boni viri prophanum" (*Adversus*, p. 2).

74. J. Du Bois, *Vaesani*, p. 134.

75. Ibid.

76. Idem, *Isagogae*, fol. 11v.

but rather of his absence of faith in authority, as was obvious from his anatomical treatise.[77] Eustachio and Du Bois were only making explicit that attitude implied by most pre-Vesalian authors on anatomical subjects, including, as we saw, Massa and Berengario. The publication of the *Fabrica,* by exacerbating the conflict between authority and evidence, between an established tradition and empirical findings, led to a radicalization of positions that brought into evidence the fideistic element behind the acceptance of the Galenic paradigm. Up to that time it had been veiled by the methodological arguments of anatomists, by their declarations of intent, by evidence of an assiduous recourse to dissection, and by a calm rhetoric; this did not mean, however, that it was not present and rigorously binding.

It may not have been accidental that the epistemological break achieved by Vesalius and the forms of the debate over ancient authority coincided temporally with the theological conflicts between Catholics and Protestants on subjects that were structurally analogous and were similar in tone. Du Bois posed the problem of the acceptance of the Galenic anatomical paradigm in terms of faith, belief, and orthodoxy, as was the case for sacred texts during those same years. Religious controversy and the climate that it generated may be the source of the language employed by Du Bois on the one hand, and Vesalian empiricism and its challenge of ancient authorities on the other. These hypothetical considerations apart, the matter of belief is important. In his book on Galenism, Owsei Temkin pointed out the risks lurking in the adoption of this issue (used in a psychological sense) to explain the lack of interest in verification that is notable in medicine as well as in other sciences based on experience (for example alchemy and natural history) during the Middle Ages and the Renaissance.[78] Belief seems pertinent, however, (and essential in the case of Du Bois) in providing a possible interpretation for the tortured relationship between text and dissection for the entire period ranging from the revival of interest in anatomy in the fourteenth century to the time of the Du Bois–Vesalius controversy. An attitude toward belief, in a variety

77. Contained in MS. C IX 17 in the Biblioteca Comunale, Siena. Its preface is transcribed in L. Belloni, "Il manoscritto senese *De dissectione et controversiis anatomicis* di Bartolomeo Eustachio," *Physis* 14 (1972): 196–97. In this work, as well as in his *De vena,* included in B. Eustachio, *Opuscula anatomica* (Venice: Vincenzo Luchinus, 1563), the author, in questions of morphology and physiology, juxtaposes the "antigrammata and "syngrammata" of the anti-Galenic anatomists, Vesalius in particular. On Eustachio, see R. French, "Natural Philosophy," pp. 453–54.

78. O. Temkin, *Galenism. Rise and Decline of a Medical Philosophy* (Ithaca and London, 1973), pp. 116–17 and note 57. On this issue, see chapter 3, "Authority and Challenge."

of ways, explains both the acceptance of a paradigm within which anatomists read, learned, thought, observed, verified and wrote, as well as the
fideistic position to which some of them resorted the moment that the
paradigm was endangered. In this sense the belief of the anatomists recalls
the cognitive resistance to change among scientists discussed first by Ludwig Fleck and subsequently by Thomas Kuhn.[79] But in contrast to this
second group the anatomists harbored beliefs that took on very definite
religious connotations and corresponded to a more general *habitus* asserted in the reception of ancient authorities and their precepts, whether
of a scientific, philosophical or theological nature.

Ludwig Fleck wrote that "Once a structurally complete and closed system of opinions consisting of many details and relations has been formed,
it offers enduring resistance to anything that contradicts it."[80] In the specific instance of anatomy, illustrated by the case of Du Bois, the belief of
the anatomist sustained and gave coherence to existing convictions, and
forcefully opposed any change. To the proofs provided by Vesalius, which
completed the process of disintegration of the Galenic paradigm that had
already begun thanks to ever more efficacious and frequent dissection, Du
Bois opposed his beliefs, which were based on the well-rooted tradition
of holding on to revealed and long-held truths. His attitude was also
strengthened by an unstated notion that the destruction of the Galenic
paradigm would have disastrous consequences for academic prestige, for
professional status, and indeed for an entire medical tradition. The paradigm now enters the social context in the way in which anatomical
knowledge was translated into the play of power within institutions. Thus
understood, the attitude of Du Bois toward tradition may be seen as the
result of cultural *habitus* and the expression of a desire to maintain a
social situation; he was not simply a prisoner of the paradigm and its
epistemological limits.

REVULSION AND UNEASE

The break in the epistemological norm that had constricted anatomy for
over a millennium because of its reliance on ancient texts, and the disintegration of the habit of belief, were inconceivable without the decisive contribution made by dissection. The spread of the practice throughout Euro-

---

79. L. Fleck, *Entstehung und Entwicklung einer wissenschaftlichen Tatsache* (Frankfurt, 1980; 1st ed. Basel, 1935); English trans., *Genesis and Development of a Scientific Fact* (Chicago, 1979); T. Kuhn, *The Structure of Scientific Revolutions* (Chicago, 1962), with a postscript in the 1970 edition.

80. L. Fleck, *Genesis*, p. 27.

pean universities in the course of the sixteenth century gradually subverted the traditional approaches to the study of the human body and, at the same time, the modes of reception of the ancient authorities. As we have seen, dissection, which was both investigative and pedagogical, had remote origins, and even early writers, especially Galen, had stressed the benefits in anatomy of a visual and manual verification on animal and human cadavers in opposition to the uncritical acceptance of what was contained in books. However, dissection was also put to improper use: it was simply auxiliary to the reading of texts, which was one of the principal reasons for the long delay before it effectively influenced the study of anatomical science. The title pages we have examined seem to illustrate clearly the changes in the application of dissection and the parallel epistemological transformations achieved by Vesalian anatomy, when at last it translated into practice those statements of principle set out by the anatomists from the days of antiquity.

We also know that Vesalius's *Fabrica* (and inevitably the *Epitome*) immediately circulated widely even outside the world of medicine and became in a few short years the principal point of reference for subsequent anatomical scholarship. However, the changes in the regulations controlling the performance of the public anatomies (such as the altered position of the instructor, who would now open the body himself and the secondary role assigned to authoritative texts), portrayed in the title page of the *Fabrica* and emphasized in many places in the volume, had no immediate influence on the organization of dissections. The annual public anatomy, the only form permitted by the university statutes, continued to be performed on the extremely limited number of subjects selected from those condemned to capital punishment, and retained substantial institutional impediments. Anatomists boasted with a growing lack of inhibition about the examinations they had performed on cadavers outside the formally permitted ones, and there were ever louder demands to view the greatest number of bodies possible. But in spite of the "Vesalian revolution" and its revision of the demonstration techniques in the anatomy lesson, as is pointed out in chapter 2, the practice of dissection remained formally constrained by a series of norms that determined how it could be carried out. These restrictions, which by now no longer corresponded to the pedagogical and epistemological requirements of sixteenth-century anatomy—especially from the time of Vesalius onward—remained in place so as to protect the study of this discipline from the criticisms and the accusations directed against those features that carried religious and anthropologically ambiguous connotations: the handling of the cadaver, the profanation and dismemberment of the bodily structure, the delayed burial. Continuing unease regarding the practice of dissection ensured that it could only take place in a context of legitimacy.

Fifteenth-and sixteenth-century treatises often carried more or less veiled allusions to the constraints that anatomists had to respect, if only formally, and of the sentiment that generated them. What might be called rhetorical strategies are the most common device used by writers of anatomical works to safeguard the legitimacy of their activity. In the majority of these volumes the preliminary pages (prefaces, letters of dedication, early chapters) carry apologies for the discipline of anatomy and a justification for the practice of dissection. This was usually argued on two points: the unavoidable necessity of studying and teaching human anatomy; its legitimacy proven through ancient tradition, and the history of anatomy itself. As we know, the principal early sources utilized for these rhetorical purposes, both before and after Vesalius, were the biological writings of Aristotle, various texts by Galen (especially the *De usu partium, Methodus medendi, the De sectis,* and the *De anatomicis administrationibus),* the *De medicina* of Celsus, and Pliny's *Naturalis Historia.*[81] Even Vesalius and his followers used the ancient authorities to justify dissection, harkening back to an established tradition and a history that had its roots in the remote beginnings of anatomical science.

Alongside the use of the ancients to construct a sphere of legitimacy for dissection, other rhetorical devices were employed for the same purpose. Vesalius's *Fabrica* provides some eloquent examples. The first is anecdotal in character. The author relates that after he had returned to Louvain from Paris he once found outside the city walls the cadaver of a man who had been hanged, whose flesh had been eaten by birds, and whose skeleton was being kept intact only by ligaments. It was the body of a thief who had been condemned to death and been left exposed to the elements for a considerable time. "I burned with such a desire and eagerness to obtain these bones," Vesalius related, so that "in the middle of the night" and "without witnesses" he remained locked outside the city walls and without experiencing any feeling of horror (*"non horruerim"*) took the cadaver apart piece by piece, brought it secretly home and, having cleaned

81. An anecdote recurring in the stories about anatomy told by anatomists (in addition to the open-minded operations of Herophilus and Erasistratus discussed thus far) is one taken from Pliny's *Naturalis Historia* (19, 26, 86) that ascribes the practice of dissection to certain Egyptian kings: "tradunt et praecordiis necessarium hunc sucum, quando phthiriasin cordi intus inhaerentem non alio potuisse depelli conpertum sit in Aegypto, regibus corpora mortuorum ad scrutandos morbos insecantibus." Among the arguments mustered to justify the discipline, just to take one example, the most frequently employed is undoubtedly Galen's. He believed anatomy could be taught and learned especially "by sight and touch" (*visu et tactu*).

the bones, reconstructed the skeleton.[82] Using this and other examples
Vesalius demonstrated the freedom with which he procured the subjects
for his observations; his familiarity with cadavers, deteriorated bodies,
and skeletons; and the total absence on his part of any religious or moral
prejudice with regard to their procurement and handling. Horror, disgust,
and shock at the sight of and even the contact with a dead body and the
procedures he carried out on it were erased for Vesalius by his yearning
for knowledge. It was this longing that impelled him to acquire a body
illicitly, outside the usual channels, thereby breaking the law. This anecdote has an illustrious antecedent in many of its details: Galen, too, in
the first book of his *De anatomicis administrationibus,* as we may recall,
tells of having appropriated the cadaver of an outlaw whose body he had
found stripped of its flesh by ravenous birds. Vesalius mentions this familiar source, emphasizing that his cadaver, just like Galen's, had been that
of a thief who had been sentenced to death, thereby spinning around his
deed a web of justifications that include ancient precedent as well as scientific motivation, and the selection for his experiments of the cadaver of
a condemned man.

The figured initials that open every chapter of the *Fabrica* represent
anatomical and medical subjects, enlivened, in the 1543 edition, by cherubs representing medical students.[83] These images convey information intended to complement the text, and like the title pages show gestures,
behavior, and aspects of the anatomical discipline that are implied or
barely mentioned in it. It may be useful to analyze briefly a few of them
that restate the problem of the procurement of cadavers. Two of these
decorative letters, the small "L" and "O" in the *De humani corporis fabrica,* depict the moment in which the body of the condemned victim, or

82. A. Vesalius, *De humani corporis fabrica,* pp. 161–62.
83. In the 1555 edition students are represented as adult men. The illuminated letters of
the *De humani corporis fabrica* were designed especially for this work. In other words, they
did not come from, as was often the case, the printer's typographical arsenal. Printers would
use the same decorated initials in several different works, even though they might bear no
relation to the subject matter of the books themselves. The letters cut for the first, 1543
edition of the *Fabrica* have been attributed to Jan Stephan von Calcar or to Domenico
Campagnola. For the second series used in the 1555 edition no attribution has been haz-
arded. See S. W. Lambert, "The Initial Letters of the Anatomical treatise: 'De Humani corporis fabrica,' of Vesalius," in S. W. Lambert, W. Wiegland, and W. M. Ivins, *Three Vesalian
Essays to Accompany the 'Icones anatomicae' of 1934* (New York, 1952), and M. Muraro
and D. Rosand, *Tiziano e la silografia veneziana del Cinquecento* (Vicenza, 1976), p. 129. The
large initials "I," "O," "Q," "T," "V" in the first edition were used again in the second.
They open each of the seven books of the *Fabrica.* The other seventeen are at the beginning
of every chapter. The initial letters of the first edition are recognizable because they are
surrounded by a double line, while those of the second are framed by only a single one. The
sole exception to this rule is the large "V," which in both editions is encased in double lines.

at least a part of it, is turned over to the students (figs. 35 and 36). In both cases this takes place in the presence of legitimizing authorities: in the letter "L" an ecclesiastical figure holding a cross in his hand is quite recognizable, while the cherubs are lowering the body of a woman from the scaffold; in the letter "O" the executioner hands down the head of the deceased in the presence of some soldiers, a metaphor for political and judicial authorities. In the large letter "I" (fig. 37) a number of cherubs are represented who are disinterring at night an apparently intact cadaver that may recently have been buried. There is nothing to suggest that it belongs to a condemned criminal. However, we do find here the symbols of a sanctioning and consenting authority: a cherub is wearing a helmet and holding a spear, another has a pennant in his hand, a third is armed with a shield. Whether these attributes were actually intended to signify the presence of authority or were designed to confuse the curious and the reader, what is interesting is that, either way, the author is resorting to an iconographic stratagem to represent the disinterment. Such a stratagem, though it may not precisely impose the presence of a guarantor, definitely suggests such a reference, even if it is couched deceptively. The "L," the "O," and the large "I" are the only three decorated letters of the twenty-three contained in the *Fabrica* in which practices related to dissection of the human body are portrayed, and the only ones in which, more or less metaphorically, civil and religious authorities make their appearance.

Another letter has something more to tell us about the status of the body to be dissected and about the anatomist/dissector. The large "V" (fig. 38) is the initial for Vesalius, and it opens the first book, the only one of the illuminated letters that depicts a mythological theme instead of a medico-anatomical subject: the story of Apollo and Marsyas. Marsyas challenged Apollo to a musical contest with the understanding that the winner would choose a penalty for the loser. The Muses were invited to judge the competition. Apollo triumphed and decreed that Marsyas should be skinned alive.[84] In the large letter "V" we see Apollo and Marsyas in the background during the musical contest, on the left the Muses in judgment, and on the right Apollo in the act of flaying his vanquished companion, who has been bound to a tree. A flute and a cloak are at the latter's feet. Three crucial moments in the myth are represented here: (1) the challenge to divinity; (2) the verdict pronounced by the

---

84. Ovid relates (*Metamorphoses*, VI: 387–91): "Clamanti cutis est summos direpta per artus/Nec quicquam nisi vulnus erat; cruor undique manat/Derectique patent nervi trepidaeque sine ulla/Pelle micant venae; salientia viscera possis/Et perlucentes numerare in pectore fibras."

Figures 35–36   Figure 35: Andreas Vesalius, the letter "L" from *De humani corporis fabrica libri septem* (Basel: Johannes Oporinus, 1543). Figure 36: Andreas Vesalius, the letter "O" from *De humani corporis fabrica libri septem* (Basel: Johannes Oporinus, 1543). In the series of decorative initials prepared for Vesalius's *Fabrica,* putti are portrayed in the act of performing procedures connected to anatomical and surgical practice. In the letter "L" two of them are retrieving the cadaver of a criminal executed by hanging so that it may be used for dissection. The majority of bodies consigned to the public anatomy lessons in Rome and in other European universities during the sixteenth century belonged to the victims of hangings. This choice was dictated by the fact that this form of execution preserved the integrity of the corporeal structure. Moreover, this type of punishment was meted out to criminals of low social rank who had been condemned for so-called "atrocious" crimes, such as theft and murder. In the letter "O" one of the putti recovers the head of a victim of beheading. Occasionally individual parts of the dead body were consigned to the anatomists. Naturally, these fragments, at the completion of the dissection, had to be interred with the rest of the condemned person's body. In the "L" as well as in the "O" the putti's activities take place under the eyes of an authority who is a guarantor of their legality. In the first letter a figure holding a cross represents religious authority; in the second soldiers symbolize civic and political authority.

Muses; and (3) the punishment. It is possible to read in this image a metaphor for anatomy expressed through a mythological allusion. Marsyas had challenged a divinity, the god of music and healing, thereby transgressing a code of conduct. The judging authorities, the Muses, condemn him and preside over his execution, which is entrusted to the hand of Apollo, who invests the roles of challenged, final judge and executioner. Marsyas thus pays a price for his daring: the violation of the satyr's body

is an execution with a connotation of contempt. Further identification can be inferred that clarifies the sense of the metaphor even more: Apollo naturally represents the anatomist and Marsyas the criminal condemned to die and to be dissected. Dissection is thus to be understood as an integral part of the punishment for the crime committed.

The shocking nature of the subject and the infamy attached to it by dissection are noted earlier in this book, when the practices connected to the anatomy lecture in the *Studium Urbis* and the special concern contained in the university's statutes to reward the soul by Masses and alms for the sacrifice of its body are considered. Alessandro Benedetti has expressed himself quite explicitly on this point. In the first chapter of his *Anatomice,* which is dedicated to the utility of anatomy and to the criteria for the selection of cadavers, after condemning the vivisection practiced in antiquity on persons who had been condemned to death as "highly ferocious" and "prohibited by our religion,"[85] he mentions that cadavers to be dissected had to belong to criminals or to those who had died from mysterious diseases. The criteria for their selection are those we have already examined: "only the ignoble, the unknown from foreign parts, can

85. See A. Benedetti, *Anatomice,* fol. 8r. The writer here harkens back, even in the choice of words, to Celsus's *De medicina,* without citing him.

Figure 37   Andreas Vesalius, the letter "I" from *De humani corporis fabrica libri septem* (Basel: Johannes Oporinus, 1543). The history of anatomy in the Renaissance and beyond is full of anecdotes concerning the theft of cadavers from cemeteries. In this decorative letter from Vesalius's *Fabrica* a group of putti is making off with a recently buried body. The violation of the grave, in this iconographic representation, takes place in the presence of other putti disguised as soldiers bearing spears, shields, helmets, and pennants.

be solicited without affront to the community and to the next-of-kin's known rights." To this Benedetti adds: "The papal constitutions have permitted this mode of dissection for a long time now; otherwise it would be held to be most execrable and abominable or against religion. There occur in addition ritual purifications of their soul and we propitiate their offenses with prayers."[86] The mention of a ritual purification to reward the soul of the deceased for the offenses inflicted on his or her remains, together with the precautions regulating the selection, give an idea of the physician's recognition of the otherworldly risks that the anatomized victim could incur as a consequence of dismemberment and profanation. This presupposes that the cadaver is endowed with some sort of sacred quality so that its dissection could be considered a sacrilegious act. This

86. Ibid, fol. 8v. He goes on: "Idque ipsi quandoque viventes in custodiis petunt, ut potius medicorum collegiis tradantur, quam carnificis manu publice trucidentur. Nec huiusmodi cadavera sine pontificum consensu attingi possunt."

Figure 38   Andreas Vesalius, the letter "V" from *De humani corporis fabrica libri septem* (Basel: Johannes Oporinus, 1543). The "V" is the first decorative letter that we encounter in Vesalius's *Fabrica*, and it is the first in the author's name. In it are depicted three moments in the myth of Apollo and Marsyas, a story that recurs frequently in anatomical iconography. Marsyas, who has become synonymous with the figure of the *écorché*, had defied the god Apollo in a musical competition and had been flayed alive in punishment. Just like the person condemned to death who ends up on the anatomist's table, Marsyas had broken a code of conduct, and the sundering of his body's integrity by Apollo (as by the anatomist) is made legitimate by the "infamous" character of the body being violated with the knife.

is a highly confused and contradictory idea, and even theologically it seems to lack foundation. The fact remains that the significance of the lasting unease revealed in the prudent strategies employed by the universities and the anatomists themselves, all of which retain strong ritualistic and religious features, is still obscure.

Other sixteenth-century evidence seems to illustrate these anxieties further, thereby giving greater substance to what has been uncovered by a close reading of the documentation. An example is offered by a case in which such misgivings are made manifest in the form of a prohibition. This occurred in Spain, where Juan de Valverde, in a letter to his protector, Cardinal Juan de Toledo, included in his *Anatomia,* makes the excuse of having written it in the vernacular because it was intended to be useful and easily comprehensible to his fellow countrymen who had never seen a cadaver dissected: "It is considered an ugly thing among Spaniards to cut up dead bodies, and also because the few who come to Italy, where they could learn about it, prefer to occupy themselves with other pastimes than this one, since they are not used to seeing such spectacles."[87] This passage makes clear the prohibition against dissection and also the disgust and revulsion that it generated, even where it was legitimate to carry it out, among those, whether physicians or students, who were not accustomed to seeing such a thing.[88]

The problem of revulsion and of familiarity as it pertains to seeing and touching dead bodies is also mentioned by Du Bois. He had two possible explanations for the age-old use of monkeys in anatomical examinations: first of all these animals were available in relative abundance; second, "they gradually accustomed diffident men to the dissection of human bodies." And he added that even among contemporary physicians, and more frequently, among young students, "several at first are initially adverse to the dissection of a human body, which causes great mental distress." He then advises them, "if they are able," first of all to try to be present at dissections and then "to get training to dissect with their own hands."[89] This "great mental distress" can be identified with those qualms and anxieties noted as a feeling of revulsion during dissection at the sight of and the contact with dead bodies as well as at the violation of their corporeal integrity. Whoever wanted to or had to know the body in all its parts in order to become a physician or anatomist, according to Du Bois, had to come to terms with this queasiness and embark on an *individual* process of familiarizing himself with the cadaver. Another expert dissector, Bartolomeo Eustachio, expressed himself in the same terms, recalling that before being able to begin learning about the human body it was essential to "remove the obstacles," and overcome the difficulties dependent on individual sensory perceptions ("the stench to the nostrils and other sensory horrors").[90]

But the document that more than all the others offers an insight into attitudes toward dissection in the sixteenth century, and not just into the difficulties that might be encountered by any young physician faced by a

87. Valverde, *La anatomia*, fol. a3r.

88. Vesalius suggests that physicians who do not dare to participate in dissections are the chief beneficiaries of the *Fabrica*'s illustrations: "si datur [bodies to be dissected], tam delicata et in medico parum probanda praediti sunt natura, ut etsi iucundissima hominis cognitione, immensi rerum Conditoris sapientiam (si quid aliud) attestante, insigniter capiantur, eo tamen animum inducere nequent, ut vel sectioni aliquando intersint" (fol. 4r).

89. J. Du Bois, *Isagogae*, fol. 93r.

90. B. Eustachio, *Anatomica Tractatio ex Vexalio et aliis Anatomicis*, Biblioteca Comunale, Siena, MS. C IX 17, fol. 2r (the preface is transcribed in L. Belloni, "Il manoscritto," p. 196).

cadaver, is undoubtedly the oration delivered by Johannes Dryander in November 1536 to inaugurate the medical courses at the University of Marburg, where he was a Protestant physician in a Protestant institution. The text is an encomium to anatomy in which, apart from the customary historical-epistemological rhetoric justifying the study of the discipline and the practice of dissection, the speaker goes on at some length about the religious and superstitious (today we would say anthropological) impediments to anatomical research of those years.[91] There were many, in fact, who "thought it disgusting that a physician and student of anatomy handle a cadaver," accusing anatomists of cruelty toward the dead because they ignored the prohibition against molesting the integrity of their remains ("injuries should not be inflicted on the dead"). "Christian piety" forbade that living criminals be destined, as had happened in the past, for the anatomical table instead of the tortures of execution, thereby wasting "innumerable opportunities advantageous to our art." If this may be comprehensible, however regretfully, wrote Dryander, there was no just and rational justification for "why superstition and every corrupt judgment of the crowd must accuse us of impiety and betray us." We should rather be grateful to the anatomist who, instead of throwing a vile cadaver to the worms, uses it for the benefit of life and of humanity.[92] The exaggerated piety and concern toward the dead, which calls for their prompt burial "in some suitable spot of ground where the duties of performing funerary rites cannot fail [to be carried out]," was also a factor that had often blocked the use of cadavers for anatomical study. Papal decrees were frequently couched in terms of such zeal regarding burial that they smacked of superstition. However, the speaker recalled, the pope himself had established "that physicians are permitted to dissect the bodies of the dead for the sake of anatomy, without adverse consequences, and with great honor." In spite of this sanction, there were still many "frightened" men who "feel such horror over the contact with dead men that they do not dare even to touch them, viewing it all with apprehension and terror."[93] Another argument used by the opponents of dissection, which is mentioned and discussed at length by Dryander, concerns "the stench of

91. *In praelectionem medicam oratio, qua Anatomiae necessarium studium commendatur, Marpurgi à Ioan. Dryandro in frequentissimo eius Academiae consessu habita. VIII Kalen. Novemb. Anno MDXXXVI*, in Dryander's *Anatomiae*, fols. aIVr–dIv. See fols. cIr–cIIIr for the defense of anatomy against superstitious notions and against the prohibitions imposed on it.

92. " . . . ex cadavere vili, mox vermibus tradendo, ad humanae vitae usum, tanta commoda, ex Anatomia cum paremus, pro tanto beneficio, qui nos odio prosequantur, haud bene gratos erga vitam hominum esse puto" (Dryander, *In praelectionem*, fol. cIIr.).

93. Ibid.

human cadavers, which was found to be intolerable during dissection and they urge total abstinence from anatomies."[94] Dryander was not, as we might have expected from a Protestant, addressing the restrictions imposed by the church and by those religious norms that regulated the administration of dead bodies. Rather, the adversaries of dissection to whom he was alluding were all those who opposed it with arguments and convictions of an anthropological nature.

This was expressed in summary fashion even in the introduction to a book published in Nuremberg in 1572 entitled *Externarum et internarum principalium humani corporis partium tabulae, atque anatomicae exercitationes, observationesque variae*. This is a work that greatly resembles what we would call today an atlas of anatomy. Its author, Folker Coïter, had studied with Gabriele Falloppio and in the second half of the sixteenth century had acquired a certain reputation as an anatomist in Italy so that he was allowed to carry out numerous dissections personally at the University of Bologna, even though he was not a member of its faculty.[95] In a section of the introduction entitled *De anatomiae utilitatibus*, he wrote:

> . . . they are to be ridiculed who vilify and cast aspersions on this art [anatomy] as being useless and unworthy of the free man. They pronounce it repugnant to handle the part of a dead person contaminated with blood and excrement. To which, I answer as follows: repugnance should be gauged by the soul, not the body; a little water can wash away the body's excrement, but the entire ocean cannot wash away the soul's ignorance. Again, they say it is cruel to cut and carve up men as a butcher would. However that may be, it is much more cruel to torture and kill the living on account of inexperience and ignorance.[96]

These texts have provided us with much food for thought. First of all they testify to the persistence of a strong resistance to dissection, even when the practice, because of the status assumed by anatomical science during the sixteenth century and because of the interventions in its favor by academic, political, and religious institutions, was recognized to have both utility and legitimacy, limited and circumscribed though these may have been. The opposition to dissection was not, as has often been main-

---

94. Ibid, fol. cIIv.

95. F. Coïter, *Externarum*. Coïter was the editor of another important anatomical work that contained Falloppio's anatomical lectures, *Lectiones Gabrielis Fallopii de partibus similaribus humani corporis—a Volchero Coiter—collectae. His accessere diversorum animalium sceletorum explicationes iconibus artificiosis et genuinis illustratae* (Nuremberg: Theodorici Gerlachii, 1575).

96. F. Coïter, *Externarum*, fol. AA2r.

tained, religious in character. It was even less (as far as public opinion was concerned) an epistemological problem, but was rather an anthropological one. The difficulty resided, it would appear, in the contact with the dead and with blood (*contrectare*) and in the desecration of the corporeal structure, "brutalize, wound, tear to shreds, lacerate" (*saevire, vulnera infligere, dilaniare, dilacerare*) all of which denote certain types of contact. The anatomist is compared to an executioner (*carnifex*), and dissection is deemed wicked (*turpis*), unworthy (*indigna*), useless (*inutilis*), cruel (*crudelis*), and vile (*foeda*). It seems clear that the accusations of impiety and cruelty directed against anatomists, as well as the arguments based on the alleged uselessness of examining cadavers, are additional indications of an anxiety secondary to the problem of the inviolability of the dead body and the risk of contamination from contact with blood and filth, as stated by Coïter. Popular superstition, Dryander would say, labeled the cadaver as unclean, and its putrefaction and the odors emanating from it supported this claim. The cadaver disgusted and repelled; contact with it, whether in its mildest form when merely handling it, or in the more fearful one of cutting into it, was unsettling and provoked horror and terror (*pavor*). If we accept what Du Bois wrote on the need to become accustomed to dissection, and what Valverde wrote concerning Spanish doctors, these sentiments of revulsion were felt even by those who would become or were already physicians. This resistance could be overcome only through a continuous and repeated examination of cadavers and an awareness of the educational, scientific, and cognitive objectives of dissection. Whoever failed to be persuaded was condemned inevitably to view the practice with horror, as dangerous and contaminating.

EPILOGUE

Dissection is not in itself an innocent act. Its practice requires, on the one hand, a legitimizing epistemological and institutional context and, on the other, the elaboration of rituals, strategies, and mechanisms to filter out its transgressive and sacrilegious connotations. A number of recent studies concerning students' reactions to the experience of dissection have shown how even within a theory of medicine that recognizes anatomy as indispensable, the opening of the cadaver can nonetheless constitute a traumatic event.[1] Finding various ways of coping with this trauma is still necessary for students today; they may for example use macabre humor or employ a highly scientific vocabulary during the demonstration. However, there are also some cultural features that characterize the teaching of contemporary medicine: in many universities dissection is still part of medical education at the clinical level, a corollary of the process of objectifying the human body that reduces the patient to the disease. This is an ethically debatable process that one encounters with a certain frequency in medical teaching, research, and practice. Precisely for the purpose of correcting this reductionism and, especially, to facilitate the encounter between medical students and the cadaver, in a number of schools, particularly in the United States, the teaching of anatomy and the practice of dissection have

1. See, for example, P. Finkelstein and L. Mathers, "Post-Traumatic Stress among Medical Students in the Anatomy Dissection Laboratory," *Clinical Anatomy* 3 (1990): 219–26; E. J. Evans and G. H. Fitzgibbon, "The Dissecting Room: Reactions of First Year Medical Students," *Clinical Anatomy* 5 (1992): 311–20; E. Godeau, "'Dans un amphithéâtre . . . :' La fréquentation des morts dans la formation des médecins," *Terrain* 20 (1993): 82–96; J. Coulehan et al., "The First Patient: Reflexions and Stories about the Anatomy Cadaver," in *Teaching and Learning in Medicine* 7 (1995): 61–66; D. R. Reifler, "'I Actually Don't Mind the Bone Saw:' Narratives of Gross Anatomy," *Literature and Medicine* 15 (1996): 183–99.

been formalized into didactic modules that take on the appearance of actual rituals. At the University of Massachusetts Medical School, for example, this teaching module includes lectures on the history of dissection, discussions about the preoccupations and the reactions arising from the experience of cutting open the cadaver, the introduction to the students of the anatomy classroom and the instruments it contains, and, in conclusion, a solemn ceremony thanking those who have voluntarily bequeathed their bodies to the department of anatomy.[2]

There is a strong temptation, naturally, to underline the continuity and persistence of certain attitudes toward dissection throughout history and the consequent elaboration of rituals intended to dissipate the qualms generated by it. But this would be historiographically suspect and certainly unproductive, and an example of facile archetypal psychologizing, were it not supported by a conspicuous effort at contextualization. What I especially want to emphasize is that qualms and unease exist even where dissection is incorporated into a solid institutional framework and an eminently coherent system of knowledge. Moreover, those aspects of dissection that could impair moral, religious, and anthropological codes (such as infraction of the unity of the body, delay in burial, and contact with blood and death) emerge especially forcefully in a context where knowledge of anatomy is not awarded an important role in the practice of medicine. As I have tried to show, beginning with Galen and persisting into the sixteenth century, this discipline was considered both a branch of natural philosophy and a field that had to be studied "on its own account," aside from its direct application to diagnosis, prognosis and healing, namely to those areas where essential medical knowledge is articulated and on which its proper social and professional justification is based. In line with the model of Rationalist medicine proposed by Galen and incorporated into university teaching, as late as the sixteenth century anatomy remained an intellectual instrument that guaranteed, intuitively and hypothetically, the logical and material foundations of medical practice, without being able to contribute to it directly since it was still bound by the constraints of the humoral paradigm.

Anatomy was more a theoretical than a practical branch of knowledge, thereby permitting medicine to be linked with philosophy. In accordance with the Galenic model, it constituted a mark of cultural, and thus also of social and professional, distinction that set those physicians endowed

2. S. Bertman and S. Marks, "The Dissecting Experience as a Laboratory for Self-Discovery about Death and Dying: Another Side of Clinical Anatomy," *Clinical Anatomy* 2 (1989): 103–13.

with university educations apart from that mass of healers—from surgeons to charlatans—who operated in the therapeutic marketplace of the ancien régime and were therefore products of an intellectually and epistemologically lower sphere. In this sense the spectacle of the public anatomy lesson solemnly repeated each year in the major European universities beginning in the fifteenth century—useless from the standpoint of anatomical research proper and, as Vesalius maintained, redundant also for educational purposes—could be understood as a celebration of the model of Galenic medicine at the high cultural level dispensed by institutions. The public anatomy lecture, as shown by the Statutes of the University of Padua in 1607, continued to be organized on the quodlibetarian model despite the pleas of Vesalius to break down the barrier separating the anatomical text from the dissection. The anatomical spectacle, in effect, made a political statement: through a display of control over anatomical knowledge the corporation of physicians publicly asserted its primacy among all the professions active in the realm of healing. It was this knowledge of anatomy that allowed physicians to justify their monopoly of such medical practices as delved within the human body (dietary prescriptions, the administration of pharmaceutical products by mouth) that had been firmly prohibited by the guild statutes to all other healers.[3]

This social and political use of anatomical knowledge proceeded in parallel to the spread of anatomy teaching and to the confirmation of dissection as a didactic and investigative tool. Nevertheless it did not strengthen the fundamentally weak status of anatomy in the educational structure of Renaissance medicine. Evidence of this can be seen in the delay with which anatomy was introduced as a basic teaching subject in the universities, in the resistance to the anatomical reforms introduced in the attempt to revise Galenic anatomy, which had supported humoral therapeutic theory, culminating in the work of Vesalius and, above all, in that complex of strategies and ritual precautions that accompanied the practice of the public dissection of the body ceremonially. The criteria by which cadavers were selected, the elaborate institutional mechanism that called for the involvement of religious, judicial, and political authorities in the administration of the dissected cadaver, as well as the norms regulating the public anatomy all over Europe must be considered expressions of those anthropological qualms caused by the desecration that the open-

---

3. Apropos the control exerted by the physicians over therapies related to the interior of the body, se G. Pomata, *La promessa di guarigione* (Bari-Rome, 1994), especially pp. 129–51; English trans., *Contracting a Cure* (Baltimore and London, 1998).

ing and the handling of cadavers involved. The purpose of dissections had not yet been sufficiently scientifically and practically justified.

The resistance to the practice of dissection and the accusations directed against anatomists in the sixteenth century were, as I point out in the final chapter, of the same sort as those voiced many centuries earlier by opponents of anatomy, as was the choice of the terms employed to articulate these sentiments. Their semantic context, from ancient Greece to Catholic and Protestant Europe at the close of the sixteenth century, is anthropological; it relates to the fear of contamination and filth connected with the handling of cadavers and with the detaching of parts of the body. The notion of foulness (*foeditas*) that recurs across the centuries is crucial in this regard. At the same time it is precisely the impurity of the decomposing body that determined funerary practices, the distancing of the cadaver from the community, and the marginalization of the trades connected with death, the carrier of contagion.[4] In the entire course of the history of anatomy the unease generated by dissection emanated, in large part, from attitudes related to the breaking of cultural and behavioral codes based on fundamentally hygienic concerns. These qualms went through different phases in the course of the centuries; they ranged from prohibitions to institutional regulations and control of the practice of anatomy, from rhetorical strategies to cautious procedures through which the handling and cutting open of cadavers was legitimized. Thus perceived, what I have defined as anxiety over dissection conceals its own anthropological basis in issues of a religious, epistemological, and social nature. These qualms are revealed in recognizable value systems on the basis of which it is possible to establish and justify what is licit or illicit.

The practices and representations of anatomy and of dissection, the norms that regulated them, and the uncertainties that preceded their acceptance in the universities of modern Europe are the product of a conflict between scientific exigencies and anthropological resistance. The criteria adopted for the selection of cadavers, unchanging from Herophilus to Vesalius and even afterwards, as well as the solemn ceremony surrounding the public anatomy lesson and the regulations covering the administration of bodies between the fifteenth and sixteenth centuries, provide the contingent forms through which anatomists and institutions responded to millenarian unease. These forms allowed public opinion, when and where justification in the name of science or of the monopoly of

---

4. See the fundamental essay by R. Hertz, "Représentation collective de la mort," *Année sociologique* (1905–1906), and E. Morin, *L'homme et la mort* (Paris, 1970), pp. 31–119.

professional qualifications did not suffice, a domesticated and reassuring version of the uses that had been made of the body. In conclusion, we are dealing with acts, choices and behavior with which, once again, human beings attempt to restrain death by means of ritual.

Geneva, November 1997

APPENDIX

Victims of Capital Punishment Dissected at Rome (1506–22 and 1556–85)

Date	Name	Origin	Crime	Sentence
3/22/12	Giovanni	Monte Lauro		Hanged
2/14/16	Caterina di Lorenzo	Corsica	Homicide	Hanged
1/1/56	Teodosio	Camerino		Hanged
1/27/57	Giovanni di Natale	Forlì		Hanged
1/13/61	Frolio d'Alí	Africa	Thefts	Hanged
1/21/61	Alessandro di Piero Santo	Spoleto	Thefts	Hanged
1/9/63	Tommè di Bonanno	Monte Fortini		Hanged
1/7/65	Antonio Blanco	France		Hanged
1/9/73	Bernardino di Giovanni	Treviso	Theft, etc.	Hanged
1/23/73	G. Domenico Sforza	Tagliacozzo		Hanged
12/18/73	Silvestro Pietrobello	Montereale	Theft, etc.	Hanged
1/15/74	Tommaso di Raffaele	Florence	Thefts	Hanged
2/5/74	Salvatore Sessa		Thefts	Hanged
1/14/75	Annibale Furlano	Parma	Theft and sodomy	Hanged
1/28/75	Paolo Buscatti	Rome	Homicide	Hanged
1/11/78	Stefano di Galeazzo	Bracciano	Homicide and theft	Hanged
2/6/79	Antonio di Lorenzo	Bergamo	Theft	Hanged
1/28/83	Anonymous			Hanged
1/20/84	Anonymous			Hanged
1/25/85	Anonymous			Hanged
1/23/86	Anonymous			Quartered

BIBLIOGRAPHY

Bibliographies are never exhaustive, and this does not pretend to be an exception. It comprises all the books and articles cited in the present text. I have added a number of works that are not directly discussed but that seem to me nevertheless to be relevant, useful, or implicitly suggested by what I have written. I have also inserted an occasional piece dealing with subjects closely related to those treated here that appeared in the interval between the publication of the Italian and English editions of this book.

With regard to the printed anatomical treatises, I have limited myself to furnishing bibliographical information, including names of publishers, on first editions and on later editions that are directly cited and discussed in the text.

This bibliography does not contain all the classical and patristic authors cited in the course of the work. Only for some cases—as, for example, for the occasional Galenic text—have I furnished the information on fifteenth- and sixteenth-century editions or modern critical editions that I have specifically described or consulted. Except where I have indicated differently in the notes, for the classics I have used the standard works. For Aristotle I relied on the edition prepared under the auspices of the Academy of Sciences of Berlin, edited by I. Bekker, 3 vols., Berlin, 1831–36 and, in particular, on the edition/translation of the *Opere biologiche*, edited by M. Vegetti and D. Lanza, Turin, 1971. For the Hippocratic writings, I refer to *Oeuvres complètes d'Hippocrate*, edited by Emile Littré, Paris, 1839–61 (L); for the works of Galen, *Claudii Galeni Opera Omnia*, edited by C. G. Kühn, Leipzig, 1821–33 (K); for other Greek physicians, *Corpus Medicorum Graecorum*, Berlin and Leipzig, 1908 ff. (CGM). The patristic texts are cited from *Patrologia Graeca*, edited by J. P. Migne, 162 vols., Paris, 1857–66 (PG) and *Patrologia Latina*, edited by J. P. Migne, 221 vols., Paris, 1844--1864 (PL).

Manuscript Collections

Archivio di Stato, Bologna
Archivio di Stato, Rome

Archivio Segreto Vaticano
Archivio di Stato, Venice
Bartholomaeus Anglicus. *Le propriétaire des choses.* Ms.fr 218, Paris: Biblio-
thèque Nationale.
Biblioteca Apostolica Vaticana
Biblioteca Nazionale Vittorio Emanuele II, Rome
Eustachio, B. *Anatomica Tractatio ex Vexalio et alijs Anatomicis.* Cod. C IX 17.
Siena: Biblioteca Comunale.
Manfredi, G. *Anathomia.* Ms. 237, Western 20287, Oxford: Bodleian Library,
1490.
Serni, P. *Trattato utilissimo per confortare i condannati a morte per via di gius-
tizia* (1665). BAV, Rome, Ms. Vat. Lat. 13596.

Printed Works
Achillini, A. *Expliciunt anatomicae [sic] annotationes Magni Alex. Achillini Bo-
non. Editae per eius fratrem Philoteum.* Bologna: Hieronymus de Benedictis,
1520.
Ademollo, A. *Le annotazioni di Mastro Titta, carnefice romano.* Città di Cas-
tello, 1886.
Agrimi, J., and C. Crisciani. *Edocere medicos. Medicina scolastica nei secoli
XIII–XV.* Naples, 1988.
Alston, M. N. "The Attitude of the Church toward Dissection before 1500." *Bul-
letin of the History of Medicine* 16 (1944): 221–38.
Andreozzi, A. "La vivisezione anatomica dei condannati a morte sotto Cosimo I
duca di Toscana." *Rivista di Discipline Carcerarie* 32 (1907): 27–33.
Annoni, J.-M., and V. Barras. "La découpé du corps humain et ses justifications
dans l'antiquité." *Canadian Bulletin for Medical History* 10 (1933): 185–227.
Aries, P. *L'homme devant la mort.* Paris, 1977.
Artelt, W. "Die ältesten Nachrichten über die Sektion menschlicher Leichen im
mittelalterlichische Abendland." *Abhandlungen zur Geschichte der Medizin
und der Naturwissenschaften* 34 (1940): 2–25.
———. "Das Titelbild zur *Fabrica* Vesals und seine kunstgeschichtlichen Voraus-
setzungen." *Centaurus* 1 (1950–51): 66–77.
Bakhtin, M. *Problems of Dostoevsky's Poetics.* Minneapolis, 1984.
———. *Rabelais and His World.* Trans. Helene Iswolsky. Cambridge, MA, 1968.
Ball, J. M. *Andreas Vesalius, the Reformer of Anatomy.* St Louis, 1910.
Barberi, F. *Il frontespizio nel libro italiano del Quattrocento e del Cinquecento.*
Milan, 1969.
Baroja, J. C. *El Carnaval. Analisis historico-cultural.* Madrid, 1965.
Bartholomaeus Anglicus. *El Libro de proprietatibus rerum.* Toulouse: Henrique
Meyer d'Alemaña, 1494.
———. *De proprietatibus rerum.* [Westminster]: Wynkyn de Worde, c. 1495.
———. *Le propriétaire des choses.* Paris: Antoine Verard, n.d.
———. *Le propriétaire des choses.* Lyons: Matthias Husz, 1482.
———. *Le propriétaire des choses.* Lyons: [Claude Davost] for Jean Dyamantier,
1500.
———. *Le propriétaire des choses.* Paris: J. Petit and M. Le Noir, 1510.

Bellarmino, R. *L'arte di ben morire*. Trans. C. Testore. Turin, 1946.

Belloni, L. "Il manoscritto senese *De dissectione et controversiis anatomicis* di Bartolomeo Eustachio." *Physis* 14 (1972): 194–200.

Benedetti, A. *Anatomice sive historia corporis humani. Ejusdem collectiones medicinales seu aphorismi.* Venice: Bernardino Guerraldo Vercellensis, 1502.

Benzi, U. (U. Senensis). *Opera.* Venice: L. A. Giunta, 1523.

Berengario da Carpi, I. *Commentaria cum amplissimis additionibus super anatomiam Mundini una cum textu ejusdem in pristinum et verum nitorem redacto.* Bologna: Hieronymum de Benedictis, 1521.

———. *Isagogae breves perlucidae ac uberrimae in anatomiam humani corporis . . . ad suorum scholasticorum preces in lucem datae.* Bologna: Benedictum Hectoris, 1522.

———. *A Short Introduction to Anatomy (Isagogae breves).* Intr. and Trans. L. R. Lind. Chicago, 1959.

Bergstrasser, G. *Neue Materialen zu Hunain ibn Ishaq's Galen-Bibliographie.* Leipzig, 1932.

Bernard, A. *La sépulture en droit canonique du Décret de Gratien au Concil de Trente.* Paris: Thèse de Droit, 1933.

Bernard, A. J. *Geoffroy Tory, peintre et graveur, premier imprimeur royal.* Paris, 1865.

Bernstein, C. M. "Titian and the Anatomy of Vesalius." *Bolletino dei Musei Civici Veneziani* 22 (1977): 39–49.

Bertman, S., and S. Marks. "The Dissecting Experience as a Laboratory for Self-Discovery about Death and Dying: Another Side of Clinical Anatomy." *Clinical Anatomy* 2 (1989): 103–13.

Biaggi, C. F. "Gli studi anatomici all'Ospedale Maggiore nel secolo XV. Leonardo a Milano." *Ospedale Maggiore* 44 (1956): 407–9.

Bliquez, L. J., and A. Kazhdan. "Four Testimonia to Human Dissection in Byzantine Times." *Bulletin of the History of Medicine* 58 (1984): 554–57.

Bonet, T. *Sepulchretum, sive anatomia practica ex cadaveribus morbo denatis.* Geneva: L. Chouet, 1679.

Bonifacius VIII. *Les registres de Boniface VIII, Bibliotheque des les Écoles Françaises d'Athènes et de Rome, 2. 4.* 4 vols. Ed. G. Digard, et al. Paris, 1904–1939.

Brockbank, W. "Old Anatomical Theatres and What Took Place Therein." *Medical History* 12 (1968): 371–84.

Brown, E. A. R. "Death and the Human Body in the Later Middle Ages." *Viator* 12 (1981): 221–70.

Brown, P. *The Cult of the Saints.* Chicago, 1981.

———. "Relics and Social Status in the Age of Gregory of Tours." In P. Brown, *Society and the Holy in Late Antiquity.* London, 1982, pp. 222–250.

Burke, P. *Popular Culture in Early Modern Europe.* London, 1978.

Busacchi, V. "Necroscopie trecentesche a scopo anatomico-patologico in Perugia." *Rivista di Storia della medicina* 9 (1965): 160–63.

Bylebyl, J. J. "Interpreting the *Fasciculo* Anatomy Scene." *Journal of the History of Medicine and Allied Sciences* 45 (1990): 285–316.

———. "The School of Padua: Humanistic Medicine in the Sixteenth Century."

In *Health, Medicine and Mortality.* Ed. C. Webster, 335–70. Cambridge and New York, 1979.

Caelius Aurelianus. *Caelius Aurelianus. On Acute and Chronic Diseases.* Ed. and trans, I. E. Drabkin. Chicago, 1950.

Campbell, D. *Arabian Medicine and Its Influence on the Middle Ages.* 2 vols. London, 1926.

Camporesi, P. *La casa dell'eternità.* Milan, 1987.

———. *Le officine dei sensi.* Milan, 1985.

Capelli, E. *La Compagnia del Neri. L'arciconfraternita dei Battuti di S.ta Maria della Croce al Tempio.* Florence, 1927.

Capparoni, P. "I maestri d'anatomia nell'Ateneo romano della Sapienza durante il sec XVI." *Bollettino dell'Istituto Storico Italiano dell'Arte Sanitaria* 6 (September–October 1926): 197–229.

———. *Memorie e documenti riguardanti Bartolomeo Eustachio pubblicati nel quarto centenario della nascita.* Fabriano, 1913.

Carafa, G. *De Gymnasio Romano et de eius Professoribus . . . libri duo.* Rome, 1751.

Carlino, A. "The Book, the Body, the Scalpel." *RES* 16 (1988): 32–50.

———. "L'exception et la règle. A propos du XVe livre du *De re anatomica* de Realdo Colombo." Ed. F.-O. Touati. *Maladies, Médecines, et Sociétés. Approches historiques pour le présent.* 2 vols. Paris, 1993, 1:170–76.

———. "Knowe Thyself. Anatomical Figures in Early Modern Europe." *RES* 27 (1995): 52–69.

Castan, P. *Naissance de la dissection anatomique: deux siècles à l'apogée du Moyen-Âge, autour d'Henri de Mondeville et Gui de Chauliac.* Montpellier, 1985.

Catalogue de la bibliothèque de feu M. le Comte Jacques Manzoni. 4 vols. Città di Castello, 1892–94.

Cavaillé, J.-P. "Un thèâtre de la science et de la mort à l'époque baroque: L'amphithêâtre de Leiden." In *Working Papers of the European University Institute.* San Domenico di Fiesole, 1990.

Cazort, M., M. Kornell, and K. B. Roberts. *The Ingenious Machine of Nature. Four Centuries of Art and Anatomy.* Ottawa, 1996.

Celsus, A. Cornelius. "A. Cornelii Celsi quae supersunt. In *Corpus Medicorum Latinorum.* Vol. 1. Ed. F. Marx. Berlin and Leipzig, 1915.

Cetto, A. M. "Die Sektion in der mittelalterlichen Miniatur." *Ciba-Symposium* 5 (1957): 168–72.

———. "Zwei unbekannte Darstellungen des Andreas Vesalius (Spicilegium anatomicum et Vesalium.)" In *Verhandlungen. XIX internationaler Kongress für Geschichte der Medizin 1964.* New York and Basel, 1966, 86–94.

Chehade, A. K. *Ibn al-Nafîs et la découverte de la circulation pulmonaire.* Damascus, 1955.

Chiffoleau, J. *La comptabilité de l'au-delà.* Rome, 1980.

Choulant, L. *History and Bibliography of Anatomic Illustration.* Ed. and trans. M. Frank. New York, 1945. (1st. ed. [German], Leipzig, 1852).

Ciardi, R. P. "Michelangelo come Galeno: Un'ipotesi iconologica." In *Studi in honore di Giulio Carlo Argan.* Ed. M. Bonicatti et al., 1:173–81. Rome, 1984.

Ciasca, R. *L'arte dei medici e speziali nella storia e nel commercio fiorentino, dal secolo XII al XV.* Florence, 1927.

Coïter, F. *Externarum et internarum principalium humani corporis partium tabulae, atque anatomicae exercitationes, observationesque variae, novis, diversis ac artificiosissimis figuris illustratae.* Nuremberg: T. Gerlach, 1572.

———. *Lectiones Gabrielis Fallopii de partibus similaribus humani corporis . . . a Volchero Coiter . . . collectae. His accessere diversorum animalium sceletorum explicationes iconibus artificiosis et genuinis illustratae.* Nuremberg: T. Gerlach, 1575.

Colombero, C. "Colombo Realdo." In *Dizionario biografico degli italiani.* Vol. 27. 241–43. Rome, 1982.

Colombo, R. *De re anatomica libri XV.* Venice: Nicolò Bevilacqua, 1559.

Condivi, A. *Vita di Michelangelo Buonarroti, pittore, scultore, architetto e gentil-huomo fiorentino pubblicata mentre viveva dal suo scolare Ascanio Condivi.* Florence, 1746. (1st. ed. Rome: Antonio Blado, 1553).

Conte, E., ed. *I maestri della Sapienza di Roma dal 1514 al 1787: i Rotuli e altre fonti.* Rome, 1991.

Coppola, E. D. "The Discovery of the Pulmonary Circulation: A New Approach." *Bulletin of the History of Medicine* 31 (1957): 47–77.

Corner, G. W. *Anatomical Texts of the Early Middle Ages.* Washington, D.C., 1927.

Corradi, A. "Dello studio e dell'insegnamento dell'anatomia in Italia nel medioevo e in parte nel Cinquecento." *Rendiconti del Regio Istituto Lombardo.* 2d ser., 15 (1873): 632–49.

Corti, M. (Mattheus Curtius). *In Mundini anatomen explicatio.* Pavia: F. Moschenus and G. B. Niger, 1550.

Coturri, E. *L'insegnamento dell'anatomia nelle università medioevali,* nona conferenza internazionale "Università e società nei secoli xii–xvi" 1979. Pistoia, 1982.

Coulehan, J., et al. "The First Patient: Reflexions and Stories about the Anatomy Cadaver." in *Teaching and Learning in Medicine* 7 (1995): 61–66.

Creutz, R. "Der Arzt Constantinus Afrikanus von Montekassino." *Studien und Mitteilungen des Benediktiner Ordnung* 47 (1929): 1–44.

Crispoldi, T. *Alcune ragioni da confortare coloro che per la giustizia publica si trovano condannati a morte.* Ancona, 1572.

Crummer, L. "Early Anatomical Fugitive Sheets." *Annals of Medical History* 5 (1923): 189–209.

———. "Further Information on Early Anatomical Fugitive Sheets." *Annals of Medical History* 7 (1925): 1–5.

———. "An Original Drawing of the Title Page of Vesalius' *Fabrica*." *Annals of Medical History.* n. s., 2 (1930): 20–30.

Cuneo, G. *Apologiae Francisci Putei pro Galeno in Anatome examen.* Venice: Franciscus de Franciscis, 1564.

Cunningham, A. *Anatomical Renaissance.* London, 1997.

Cushing, H. W. *A Bio-Bibliography of Andreas Vesalius.* New York, 1943.

Dall'Osso, E. *L'organizzazione medico legale a Bologna e a Venezia nei secoli XII–XIV.* Cesena, 1956.

D'Alverny, M.-T. "Pietro d'Abano traducteur de Galien." *Medioevo* 11 (1985): 19–64.

D'Angelo, B. *Ricordo del ben morire.* Brescia, 1589.

Darrouzès, J., ed. Symeon le Nouveau Théologien. *Traités théologiques et éthiques.* 2 vols. Paris, 1966–67.

Dawson, W. R. "Making a Mummy." *Journal of Aegyptian Archeology.* 13 (1927): 40–49.

Deichgraber, K. *Die griechische Empirikerschule: Sammlung der Fragmente und Darstellung der Lehre.* Berlin and Zurich, 1965. (1st. ed. 1930).

Del Gaizo, M. *Dell'azione dei Papi sul progresso dell'anatomia e della chirurgia sino al 1600.* Milan, 1893.

———. *Sulla pratica dell' anatomia in Italia sino al 1600.* Naples, 1892.

De Lint, J. G. "Fugitive Anatomical Sheets." *Janus* 28 (1924): 78–91.

Del Re, N. *Monsignor Governatore di Roma.* Rome, 1972.

De Maio, R. *Michelangelo e la Controriforma.* Bari and Rome, 1978.

De Sandre Gasparini, G. "La confraternita di San Giovanni Evangelista della Morte in Padova e una 'riforma' ispirata dal vescovo Barozzi." In *Miscellanea G. G. Meersseman.* Vol. 2. 765–815. Padua, 1970.

Di Benedetto, V. *Il medico e la malattia. La scienza di Ippocrate.* Turin, 1986.

Dinsmoor, W. B. "The Literary Remains of Sebastiano Serlio." *The Art Bulletin* 24 (1942): 55–91.

Di Pietro, P. "Contributo alla storia degli studi anatomici in Modena." In *Atti e memorie della Deputazione di Storia Patria per le Antiche Provincie Modenesi.* Series 8, Vol. 9. 1957, pp. 81–87.

Doherty, D. *The Sexual Doctrine of Cardinal Cajetan.* Studien zur Geschichte der katholischen Moraltheologie, Vol 12. Regensburg, 1966.

Douglas, J. *Bibliographiae anatomicae specimen, sive catalogus omnium pene auctorum qui ab Hippocrate ad Harveum rem anatomicam . . . scriptis illustrarunt.* London, 1715.

Douglas, M. *Natural Symbols.* Harmondsworth, 1970.

———. "Pollution." In *International Encyclopedia of the Social Sciences.* Vol. 12, 336–42. New York, 1968.

———. *Purity and Danger. An Analysis of Concepts of Pollution and Taboo.* London, 1966.

Dryander, J. *Anatomia capitis.* Marburg: E. Cervicornus, 1536.

———. *Anatomiae, hoc est, corporis humani dissectionis pars prior, in qua singula quae ad caput spectant recensentur membra . . . Item anatomia porci, ex traditione Cophonis, infantis, ex Gabriele de Zerbis.* Marburg: E. Cervicornus, 1537.

———. "In praelectionem medicam oratio, qua Anatomiae necessarium studium commendatur, Marpurgi à Ioan. Dryandro in frequentissimo eius Academiae consessu habita. VIII Kalen. Novemb. Anno MDXXXVI." In J. Dryander, *Anatomiae, hoc est, corporis humani dissectionis pars prior, in qua singula quae ad caput spectant recensentur membra . . . Item anatomia porci, ex traditione Cophonis, infantis, ex Gabriele de Zerbis.* Marburg: E. Cervicornus, 1537.

Du Bois, J. (Iacobus Sylvius). *In Hippocratis et Galeni physiologiae partem anatomicam isagogae.* Paris: J. Hulpeau, 1555. (1st ed. Paris, 1542).

———. *Institutiones anatomicae. In quibus ea omnia quae ad rem anatomicam bene docendam pertinere possunt, brevi ac facili modo ostenduntur.* Venice, 1572.

———. *Opera medica jam demum in sex partes disgesta, castigata et indicibus necessariis instructa.* Geneva: Chouet, 1630.

———. *Ordo et ordinis ratio in legendis Hippocratis et Galeni libris.* Paris: apud Aegidium Gorbinum, 1561. (1st ed. Paris: ex off C. Wecheli, 1539).

———. *Vaesani cuiusdam calumniarum in Hippocratis Galenique rem anatomicam depulsio.* Paris, 1552. Reprinted in R. Henerus, *Adversus Iacobi Sylvii depulsionum anatomicarum calumnias, pro Andrea Vesalio Apologia.* Venice: [Gualtiero Scoto], 1555, 71–134.

Duden, B. *The Woman Beneath the Skin. A Doctor's Patients in XVIIIth Century Germany.* Cambridge, MA and London, 1991. (1st ed. [German] Stuttgart, 1987.)

Dulieu, L. *La médecine à Montpellier.* 6 vols. in 9 tomes. Avignon, 1975–94.

Dürer, A. *Hierinn sind begriffen vier Bücher von menschlicher Proportion.* Nuremberg, 1528.

Durling, R. J. "A Chronological Census of Renaissance Editions and Translations of Galen." *Journal of the Warburg and Courtauld Institutes* 24 (1961): 230–305.

Edelstein, L. "The Development of Greek Anatomy." *Bulletin of the Institute of the History of Medicine* 3 (1935): 235–48.

———. "The History of Anatomy in Antiquity." In *Ancient Medicine.* Ed. O. Temkin and C. L. Temkin, 247–301. Baltimore, MD. 1967. (Originally, "Die Geschichte der Sektion in der Antike." *Quellen und Studien zur Geschichte der Naturwissenschaften und der Medizin* 3, no. 2 [1932–33]).

Edgerton, S. Y., Jr. *Pictures and Punishment. Art and Criminal Prosecution during the Florentine Renaissance.* Ithaca and London, 1985.

Ermini, G. *Storia dell'Università di Perugia.* Bologna, 1947.

Estienne, C. (Carolus Stephanus). *De dissectione partium corporis, una cum figuris et incisionum declarationibus a Stephano Riverio expositis.* Paris: apud Simonem Colinaeum, 1545.

———. *Opuscula anatomica.* Venice: Vincenzo Luchino, 1563.

———. *Tabulae anatomicae Bartholomaei Eustachii, quas e tenebris tandem vindicatas . . . praefatione, notisque illustravit ac publici juris fecit Jo. Maria Lancisius.* Rome, 1714.

Evans, E. J., and G. H. Fitzgibbon. "The Dissecting Room: Reactions of First Year Medical Students." *Clinical Anatomy* 5 (1992): 311–20.

Fabroni, A. *Historiae Academiae Pisanae.* 3 vols. Pisa, 1791–95.

Facciolati, J. *Fasti Gymnasii Patavini.* Padua, 1757.

———. *De Gymnasio Patavino Syntagmata XII.* Padua, 1752.

Farinacci, P. *Praxis et theoria criminalis.* 3d ed. Venice, 1603.

Ferrari, G. "Public Anatomy Lessons and the Carnival: the Anatomy Theatre of Bologna." *Past and Present* 117 (1987): 50–106.

Feyerabend, P. *Against Method.* London, 1975.

Finkelstein. P., and L. Mathers."Post-Traumatic Stress among Medical Students in the Anatomy Dissection Laboratory." *Clinical Anatomy* 3 (1990): 219–26.

Firpo, L. "Esecuzioni capitali in Roma (1567–1671)." In *Eresia e Riforma nell'Italia del Cinquecento,* 307–42. Florence and Chicago, 1974.

Fleck, L. *Entstehung und Entwicklung einer wissenschaftlichen Tatsache.* Frankfurt, 1980. (1st ed. Basel, 1935.) English trans., *Genesis and Development of a Scientific Fact.* Chicago, 1979.

Foucault, M. *L'archéologie du savoir.* Paris, 1969.

———. *Le souci de soi.* Paris, 1984.

Fournier, M. *Les Statuts et privilèges des universités françaises depuis leur fondation jusqu'en 1789.* 4 vols. Paris, 1890–94.

Fraser, P. M. "The career of Erasistratus of Ceos." *Rendiconti dell'Istituto Lombardo. Classe di Lettere e Scienze Morali e Storiche* 103 (1969): 518–37.

———. *Ptolemaic Alexandria.* Oxford, 1986. 3 vols. (1st ed. Oxford, 1972).

French, R. "Berengario da Carpi and the Use of Commentary in Anatomical Teaching." In *The Medical Renaissance of the Sixteenth Century.* Ed. A. Wear, et al. Cambridge, 1985, 42–74.

———. "A Note on the Anatomical Accessus of the Middle Ages." *Medical History* 23 (1979): 426–63.

French, R., and G. E. R. Lloyd. "'De juvamentis membrorum' and the Reception of Galenic Physiological Anatomy." *Isis* 70 (1970): 96–109.

———. "Lost Greek Fragments of Galen's Anatomical Procedures," *Sudhoff Archiv* 62 (1978): 235–249.

———. "Natural Philosophy and Anatomy." In *Le corps à la Renaissance. Actes du XXXe colloque de Tours, 1987.* Ed. J. Céard, M.-M. Fontaine, and J.-C. Margolin, Paris, 1990, 447–60.

———. "A note on the Anatomical Accessus of the Middle Ages." *Medical History* 23 (1979): 426–63.

Freud, S. *Totem and Taboo.* New York, 1950. (1st ed. [German] Leipzig and Vienna, 1913).

Froriep, L. F. *Über die anatomischen Anstalten zu Tübingen.* Weimar, 1811.

Fuchs, L. *De humani corporis fabrica ex Galeni et Andreae Vesali libris concinnatae, epitomes pars prima.* Lyons: I. Frellonium, 1551–55.

Fulgentius, F. P. *Opera.* Ed. R. Helm. Lipsiae, 1898.

Galen. *De anatomicis administrationibus. De constitutione artis medicinae. De theriaca ad Pisonem. De pulsibus ad medicinae candidatos liber. Per Joan. Guinterium Andernacum Latinitate jam recens donata.* Basel: A. Cratander, 1531.

———. *De anatomicis administrationibus libri novem Joanne Guinterio Andernaco Medico interprete.* [Paris]: S. Colinaeus, 1531.

———. "De libris propriis." In Galen, *Scripta minora.* Vol. 2. Ed. I. Müeller. Leipzig, 1891, pp. 104–108.

———. *Galen on Anatomical Procedures.* Intr. and trans. C. Singer. Oxford and London, 1956.

———. *Galen on the Usefulness of the Parts of the Body.* Intr. and trans. M. T. May. Ithaca and London, 1968.

———. *Libri Anatomici.* Bologna: G. B. Phaelli, 1529.

———. *Methodus medendi.* Paris: S. de Colines, 1530.

———. *On Anatomical Procedure. The Later Books.* Trans. D. E. Duckworth. Cambridge, 1962.

———. *Opera Omnia.* Pavia: J. de Burgofranco, 1515–16.

———. *Operum impressio novissima . . . omnes tam veteres quam novas interpretationes continet. . . .* Venice: L. A. Giunta, 1528.

———. *Procedimenti anatomici.* Intr. and trans. I. Garofalo. 3 vols. Milan, 1991.

————. *Therapeutica.* Venice: L. A. Giunta, 1522.

Garcia Ballester, L. *Galeno, en la sociedad y en la ciencia de su tiempo (c. 130–200 d-de C).* Madrid, 1972.

————. "El saber anatómico de Praxagoras de Cos." *Bolletino de la Sociedad Española de Historia Medica* 7 (1966): 43–49.

Garofalo, I. *Erasistrati Fragmenta,* Pisa, 1988.

Gelis, J., and O. Redon, eds. *Les miracles miroirs des corps.* Paris, 1983.

Geminus, T. *Compendiosa totius anatomie delineatio.* . . . London: in off. J. Herfordie, 1545.

Gherardi, A., ed. *Statuti dell'Università e Studio di Firenze.* Florence, 1881.

Godeau, E. "'Dans un amphithéâtre . . . :' La fréquentation des morts dans la formation des médecins." *Terrain* 20 (1993): 82–96.

Gould, C. "Sebastiano Serlio and Venetian Painting." *Journal of the Warburg and Courtauld Institutes* 25 (1962): 58–64.

Gregorio Magno. *Dialogi.* Ed. U. Moricca. Rome, 1924.

Guerra, F. "The Identity of the Artists Involved in Vesalius' *Fabrica* 1543." *Medical History* 13 (1969): 37–50.

————. "Juan de Valverde de Hamusco." *Clio Medica* 2 (1968): 339–63.

Gugliemo da Saliceto. *Liber magistri Gulielmi placentini de Saliceto in scientia medicinali . . . qui Summa conservationis et curationis appellatur. Item Cyrurgia.* Venice: [D. Berthocus or M. Saraceno], 1490.

Guido de Cauliaco. *Ars chirurgica Guidonis Caulici.* Venice: Giunta, 1546.

————. *Cyrurgia Guidonis de Cauliaco et Cirugia Bruni, Theodorici, Rolandi, Rogerii, Bertapaliae, Lanfranci.* Venice: cura Boneti Locatelli, mandato Octaviani Scoti, 1498.

————. *Le livre appelé Guidon. De la pratique en chirurgie.* Lyons: Jehan de Vingle, 1498.

Guillaume d'Auvergne. *Opera Omnia.* Paris, 1674.

Haller, A. von. *Bibliotheca anatomica, qua scripta ad anatomen et physiologiam facientia a rerum initiis recensentur.* 2 vols. Zurich, 1774–77. Reprint, Heidelsheim and New York, 1969.

Harcourt, G. "Andreas Vesalius and the Anatomy of Antique Sculpture." *Representations* 17 (1987): 28–61.

Harig, G., and J. Kollesch. "Galen und Hippokrates." In *La collection hippocratique et son rôle dans l'histoire de la médecine.* Ed. L. Bourgey and J. Jouanna. Leiden, 1975, 257–74.

Harris, C. R. S. *The Heart and the Vascular System in Ancient Greek Medicine from Alcmaeon to Galen.* Oxford, 1973.

Heckscher, W. S. *Rembrandt's Anatomy of Dr. Nicolaas Tulp: An Iconological Study.* New York, 1958.

Henerus, R. *Adversus Iacobi Sylvii depulsionum anatomicarum calumnias, pro Andrea Vesalio Apologia.* Venice: [Gualtiero Scoto], 1555.

Herrlinger, R. *A History of Medical Illustration from Antiquity to 1600 A.D.* London, 1970. (1st ed. [German], Munich, 1967).

Hertz, R. "Représentation collective de la mort." In *Sociologie religieuse et folklore.* Paris, 1970, 1–83. Also in *Année Sociologique,* 1st ser., 10 (1907).

Heseler, B. *Andreas Vesalius' First Public Anatomy at Bologna, 1540.* Ed. R. Eriksson. Uppsala, 1959.

Hind, A. M. *An Introduction to a History of Woodcut.* 2 vols. London, 1935.

Hippocrates. *Opere.* Trans. M. Vegetti. Turin, 1964.

Höllander, E. *Die Medizin in der klassischen Malerei.* Stuttgart, 1923.

Huillard-Breholles, J. L. A. *Historia diplomatica Friderici Secundi.* 12 vols. Paris, 1852–61.

Hunain Ibn Ishaq. *Über die syrischen und arabischen Galen Übersetzungen.* Ed. G. Bergstrasser. Leipzig, 1925.

Hutten, U. von. *De guaiaci medicina et morbo gallico liber unus.* Venice: Zilettus, 1567.

Hyatt Mayor, A. *Artists and Anatomists.* New York, 1984.

Ilberg, J. "Uber die Schriftstellerei des Klaudius Galenus." *Rheinisches Museum für Philologie* 44 (1889), pp. 207–39; (1892): 489–514; (1896) 165–96; (1897): 591–623.

Isnardi, P. L. *Storia dell'Università di Genova.* 2 pts. Genoa, 1861–67.

Jacquart, D., and R. Micheau. *La médecine arabe et l'Occident médiéval.* Paris, 1990.

Jaeger, W. *Diokles von Karystos: Die griechische Medizin und die Schule des Aristoteles.* Berlin, 1938.

Janson, H. W. *Apes and Ape Lore in the Middle Ages and the Renaissance.* London, 1952.

———. "Titian's Laocoon Caricature and the Vesalian-Galenist Controversy." *The Art Bulletin* 28 (1946): 49–56.

John of Ketham. *Fasciculus medicinae.* Milan: G. da Castiglione, 1509.

———. *Fasciculo di medicina.* Venice: G. and G. de Gregoriis, 1493.

———. *Fasciculus medicinae.* Venice: G. and G. de Gregoriis, 1491.

John of Salisbury. *Metalogicus.* Ed. C. J. Webb. Oxford, 1929.

Jones, W. H. S. "Introduction." *Hippocrates.* 4 vols. 1923–31. London and Cambridge, MA: Loeb Classical Library, 1:i–xlix.

Karp, D. R, et al., eds. *Ars Medica. Art, Medicine and the Human Condition.* Philadelphia, 1985.

Keller, A., ed. *A Hangman's Diary: Being the Journal of Master Franz Schmidt, Public Executioner of Nuremberg, 1573–1617.* London, 1928.

Kellett, C. E. "Perino del Vaga et les illustrations pour l'anatomie d'Estienne." *Aesculape* 37 (1955): 74–89.

———. "Sylvius and Reform of Anatomy." *Medical History* 5 (1961): 101–16.

Kemp, M. "A Drawing for the *Fabrica,* and Some Thoughts upon the Vesalius Muscle-Men." *Medical History* 14 (1970): 277–88.

Kristeller, P. O. "The School of Salerno: Its Development and Its Contribution to the History of Learning." *Bulletin of the History of Medicine* 17 (1945): 495–551. (Italian trans.: "La Scuola di Salerno: il suo sviluppo e il suo contributo alla storia della scienza." In P. O. Kristeller, *Studi sulla scuola medica salernitana,* Naples, 1986, 11–96.

Kuhlewein, H. *Die chirurgischen Schriften des Hippokrates.* Ilfeld, 1897–98.

Kuhn, T. "Second Thoughts on Paradigms." 293–319. In T. Kuhn, *The Essential Tension.* Chicago, 1977.

———. *The Structure of Scientific Revolutions.* Chicago, 1962, 1970.

Kurz, O. "Medical Illustrations (2): the Ketham Group." *Journal of the Warburg and Courtauld Institutes* 5 (1942): 138–41.

Kusukawa, S. *The Transformation of Natural Philosophy: The Case of Philip Melanchthon.* Cambridge and New York, 1995.

Laguna, A. de. *Anatomica methodus, seu de sectione corporis contemplatio.* Paris: J. Kerver, 1535.

Lambert, S. W., W. Wiegland, and W. M. Ivins. *Three Vesalian Essays to Accompany the Icones Anatomicae of 1934.* New York, 1952.

Laqueur, T. "Amor Veneris, vel Dulcedo Appeletur." In *Fragments for a History of the Human Body.* Ed. M. Feher, et al. New York: Zone 5, 1989, pt. 3, 90–131.

———. *Making Sex: Body and Gender from the Greeks to Freud.* Cambridge, MA and London, 1990.

Lassek, A. M. *Human Dissection: Its Drama and Struggle.* Springfield, IL, 1958.

Le Goff, J. *La naissance du Purgatoire.* Paris, 1981.

Lind, L. R. *'The Epitome' of Andreas Vesalius.* Cambridge, 1949.

———. *Studies in Pre-Vesalian Anatomy: Biographies, Translations, Documents.* Philadelphia, 1975.

Lindberg, D.C. "The Transmission of Greek and Arabic Learning to the West." In *Science in the Middle Ages.* Ed. D.C. Lindberg. Chicago and London, 1978, 52–90.

Linebaugh, P. "The Tyburn Riot against the Surgeons," in *Albion's Fatal Tree.* Ed. D. Hay. London, 1975, pp. 65–117.

Lloyd, G. E. R. "Alcmaeon and the Early History of Dissection." In *Methods and Problems in Greek Science,* 164–93. Cambridge and New York, 1991 (originally published in *Sudhoff Archiv* 59 (1975): 113–47).

———. *Magic, Reason and Experience: Studies in the Origin and Development of Greek Science.* Cambridge, 1979.

———. *Polarity and Analogy: Two Types of Argumentation in Early Greek Thought.* Cambridge, 1966.

———. *Science, Folklore and Ideology: Studies in the Life Sciences in Ancient Greece.* Cambridge, 1983.

Longrigg, J. "Anatomy in Alexandria in the Third Century B.C." *British Journal for the History of Science* 21 (1988): 455–88.

Malagola, C., ed. *Statuti dell' Università e dei Collegi dello Studio bolognese.* Bologna, 1888.

Manara, G. *Notti malinconiche. Nelle quali con occasione di assister' à condannati a morte, si propongono varie difficoltà spettanti a simile materia. Serviranno per istruttione à confessori, confortatori, e altri assistenti nelle conforterie. . . .* Bologna, 1668.

Mandosius, P. *Theatron in quo Maximorum Christiani Orbis Pontificum Archiatros . . . spectandos exhibet.* Rome, 1696.

Marini, G. L. *Degli Archiatri Pontifici.* 2 vols. Rome, 1784.

Martinotti, G. "L'insegnamento dell'anatomia in Bologna prima del secolo XIX." *Studi e memorie per la storia dell'università di Bologna* 2 (1911): 3–146.

Massa, N. *Liber introductorius anatomiae sive dissectionis corporis humani, nunc primum ab ipso auctore in luce aeditus in quo quamplurima membra operationes et utilitates tam ab antiquis quam a modernis praetermissa manifestantur.* Venice: Francisci Bindoni and Maphei Pasini, 1536.

McDermott, W. C. *The Ape in Antiquity.* Baltimore, 1938.

McVaugh, M. R., and N. Siraisi, eds. "Renaissance Medical Learning. Evolution of a Tradition." *Osiris,* 2d. ser., 6 (1990).

Medici, M. *Compendio storico della scuòla anatomica di Bologna.* Bologna, 1857.

Medici, Z. *Trattato utilissimo di conforto de' condannati a morte per via di giustizia.* Ancona, 1572.

Merton, R. K. *The Sociology of Science.* Chicago, 1973.

Meyer, A. W., and S. K. Wirt. "The Amuscan Illustration." *Bulletin of the History of Medicine* 15 (1943): 667–87.

Meyer-Steineg, T., and K. Sudhoff. *Geschichte der Medizin im Überblick.* Jena, 1928.

Meyerhof, M. "Ibn al Nafîs (13th cent.) and His History of Lesser Circulation." *Isis* 23 (1935): 100–120.

———. "New Light on Hunain ibn Ishaq and His Period." *Isis* 14 (1926): 685–724.

Michler, M. "Guy de Chauliac als Anatom." In *Frühe Anatomie.* Ed. R. Herrlinger and F. Kudlien. Suttgart, 1967, 15–32.

Middendorpius, I. *Academicarum Orbis Christiani.* N. p., 1583.

Moes, R. J., and C. D. O'Malley. "R. C.: 'On Those Things Rarely Found in Anatomy.'" *Bulletin of the History of Medicine* 34 (1960): 508–28.

Mondeville, H. de. *Cyrurgia.* Ed. J.-L. Pagel. Berlin, 1892.

Mondino dei Liuzzi. *Anathomia.* Padua: Petrus Maufer, c. 1474.

———. *Anathomia Mundini. Emendavit Martinus Mellerstat.* Leipzig: M. Landsberg, 1493.

———. *Anatomia.* Geneva, 1519.

———. *Anatomia.* Paris: A. Lotrian and D. Janot, 1532.

———. *Anatomia Mundini.* Marburg: C. Egenolphus, 1541.

———. *Anatomia, riprodotta da un codice bolognese del secolo XIV e volgarizzata nel secolo XV.* Ed. L. Sighinolfi. Bologna, 1930.

———. *De omnibus corporis humani membris interioribus anatomia cum figuris faberrimis non solum medicis sed philosophantibus etiam omnibus utilissima.* Rostock: Nikolaus Marschalk, 1514.

Montaigne, M. de. *Journal de voyage en Italie par la Suisse et l'Allemagne 1580–81.* Paris, 1946.

Morin, E. *L'homme et la mort.* Paris, 1970.

Morland, H. "Die lateinischen Oribasius-Übersetzungen." In *Symbolae Osloenses.* Fasc. V. Oslo, 1932.

Mudry, P. *La préface du 'De Medicina' de Celse.* Rome, 1982.

Muhammed ibn Zakariyya al-Razi, 'Ali ibn al-'Abbas, and 'Ali ibn Sina. *Trois traités d'anatomie arabe.* Trans. P. De Konig. Leiden, 1903.

Münster, L. "La medicina legale in Bologna dai suoi albori alla fine del XIV secolo (nota preventiva)." *Bollettino dell'Accademia Medica F. Pacini* 26 (1955): 257–71.

Muraro, M., and D. Rosand. *Tiziano e la silografia veneziana del Cinquecento.* Vicenza, 1976.

Muraro, M. "Tiziano e le anatomie di Vesalio." In *Tiziano e Venezia: convegno internazionale di studi (1976).* Vicenza, 1980, 307–316.

Musso, C. *Prediche.* Venice, 1573.

Nardi, M. "Statuti e documenti riflettenti la dissezione anatomica umana e la nomina di alcuni lettori di medicina nell'antico studium generale fiorentino." *Rivista di storia delle scienze mediche e naturali* 47 (1956): 237–49.

Nasr, S. H. *Science and Civilization in Islam.* New York, 1968.

Neuberger, M. *Geschichte der Medizin.* Stuttgart, 1906.

Nutton, V. K. G. *Kühn and his Edition of the Works of Galen.* Oxford, 1976.

O'Malley, C. D. *Andreas Vesalius of Brussels, 1514–64.* Berkeley and Los Angeles, 1964.

Ongaro, G. "La medicina nello studio di Padova e nel Veneto." In *Storia della cultura veneta.* Vol. 3. Ed. G. Arnaldi and M. Pastore Stocchi. Pt. 3, 95–134. Vicenza, 1981.

Opelt, I. "Zur Übersetzungstechnik des Gerhard von Cremona." *Glotta* (1960): 135–70.

Ortalli, G. "La perizia medica a Bologna nei sec. XIII e XIV. Normativa e pratica di un istituto giudiziario." *Atti e memorie della Deputazione di Storia Patria per le provincie di Romagna.* n. s. 17–19 (1965–68): 223–59.

Ottosson, P.-G. *Scholastic Medicine and Philosophy. A Study on Galen's 'Tegni' (ca. 1300–1450).* Naples, 1984.

Pagel, W. *William Harvey's Biological Ideas.* Basel and New York, 1966.

Paglia, V. *La morte confortata. Riti della paura e mentalità religiosa a Roma nell'età moderna.* Rome, 1982.

Panzani, G. *De venetae anatomes historia et claris Venetiarum anatomicis prolusio.* Venice, 1763.

Paravicini Bagliani, A. *I testamenti dei Cardinali del Duecento.* Rome, 1980.

———. "Storia della scienza e storia della mentalità. Ruggero Bacone, Bonifacio VIII e la teoria della 'prolungatio vitae.'" In *Aspetti della letteratura latina del secolo XIII. Atti del primo convegno internazionale di studi dell'Associazione per il Medioevo e l'umanesimo latini (AMUL). Perugia 3–5 ottobre 1983.* Ed. C. Leonardi and G. Orlandi, 243–80. Perugia and Florence, 1986.

Park, K. "The Criminal and Saintly Body: Autopsy and Dissection in Renaissance Italy." *Renaissance Quarterly* 47 (1994): 1–33.

———. *Doctors and Medicine in Early Renaissance Florence.* Princeton, 1985.

———. "The Life of the Corpse: Division and Dissection in Late Medieval Europe." *Journal of the History of Medicine and Allied Sciences* 50 (1995): 111–32.

Parronchi, A. "Michelangelo e Realdo Colombo." In *Opere giovanili di Michelangelo.* Vol. 2, 191–233; vol. 3, 159–66. Florence, 1968–92.

Petrioli, G. *Corso anatomico o sia Universal commento nelle tavole del celebre Bartolomeo Eustachio.* Rome, 1742.

———. *Le otto tavole anatomiche con cinquanta figure in foglio delineate per compimento dell'opera sublime et imperfetta del celebre Bartolomeo Eustachio.* Rome, 1750.

Petrus, de Abano. *Conciliator controversiarum, quae inter philosophos et medicos versantur.* Venice: Heirs of L. A. Giunta, 1565.

Piazza, B. *Eusevologio romano overo delle opere pie di Roma . . . con due trattati delle Accademie e Librerie celebri di Roma.* Rome, 1698.

Piccolomini, A. *Anatomicae prelectiones. . . .* Rome: B. Bonfadini and T. Diani, 1586.

Pigeaud, J. "Formes et normes dans le *De Fabrica* de Vésale." In *Le corps à la Renaissance. Actes du XXXe colloque de Tours, 1987.* Ed. J. Céard, M.-M. Fontaine, and J.-C Margolin, 399–421. Paris, 1990.

Platter, F. *De corporis humani structura et usu libri III . . . Tabulis explicati iconibus . . . illustrati.* Basel: Froben, 1581.

Polverini Fosi, I. "Pietà, devozione e politica: due confraternite fiorentine nella Roma del Rinascimento." *Archivio Storico Italiano* 149 (1991): 119–161.

Popper, K. *The Logic of Scientific Discovery.* London, 1959.

Portal, A. *Histoire de l'anatomie et de la chirurgie.* 6 vols. Paris, 1770–73.

Pouchelle, M. C. *Corps et chirurgie à l'apogée du Moyen Age.* Paris, 1983.

Pozzi, F. (Franciscus Puteus). *Apologia in anatome pro Galeno, contra Andream Vesalium Bruxellensis.* Venice: F. de Portonariis, 1562.

Premuda, L. *Storia dell'iconografia anatomica.* Milan, 1957.

Prosperi, A. "Il sangue e l'anima: Ricerche sulle compagnie di giustizia in Italia." *Quaderni storici* 15, no. 51 (1982): 959–99.

Putti, V. *Berengario da Carpi.* Bologna, 1937.

Rath, G. "Pre-Vesalian Anatomy in the Light of Modern Research." *Bulletin of the History of Medicine* 35 (1961): 142–48.

Reifler, D. R. "'I Actually Don't Mind the Bone Saw:' Narratives of Gross Anatomy." *Literature and Medicine* 15 (1996): 183–99.

Renazzi, F. M. *Storia dell'Università degli Studi di Roma, detta comunemente la Sapienza.* Rome, 1804.

Renouard, P. *Bibliographie des éditions de Simon de Colines 1520–40.* Paris, 1894.

Richardson, R. *Death, Dissection and the Destitute.* London, 1987.

Richardus. *Die Anatomie des Magister Richardus.* Ed. J. Florian. Breslau, 1875.

Roberts, K. B., and J. D. W. Tomlinson. *The Fabric of the Body: European Traditions of Anatomical Illustration.* Oxford, 1992.

Rossi, P. *I filosofi e le macchine (1400–1700).* Milan, 1976.

Roth, M. *Andreas Vesalius Bruxellensis.* Berlin, 1892.

Rufus of Ephesus. *Oeuvres de Rufus d'Ephèse. Textes collectionnés sur les manuscrits, traduits pour la première fois en français.* Ed. C. Daremberg and E. Ruelle. Paris, 1879.

Russell, K. F. *The "De re anatomica" of Realdus Colombus.* Melbourne, 1953.

Salimbene de Adam. *Cronica.* Vol. 2. Ed. F. Bernini. Bari, 1942.

Sander, M. *Le livre à figures italien depuis 1467 jusqu'à 1530.* 6 vols. Milan, 1942–43.

Santi, F. "Il cadavere e Bonifacio VIII, tra Stefano di Tempier e Avicenna. Intorno a un saggio di Elisabeth Brown." *Studi Medievali.* 3d ser., 28 (1987): 861–78.

Santorio, S. *Commentaria in primam fen primi libri Canonis Avicennae.* Venice: I. Sarcinam, 1626.

Sarton, G. *Introduction to the History of Science.* 3 vols. Washington, D.C., 1927–48.

Saunders, J. B. de C. M., and C. D. O'Malley. *The Anatomical Drawings of Andreas Vesalius.* New York, 1950.

———. *The Illustrations from the Works of Andreas Vesalius of Brussels.* Cleveland, 1950.

Sawday, J. *The Body Emblazoned. Dissection and the Human Body in Renaissance Culture.* London and New York, 1995.

Scanarolo, G. B. *De visitatione carceratorum libri tres.* Rome, 1655.

Scarborough, J. "Celsus on Human Vivisection in Ptolemaic Alexandria." *Clio Medica* 11 (1976): 25–38.

Schipper, G. "Ein neu text der gynaecia des Vindicianus." Med. diss., University of Leipzig, 1921.

Schipperges, H. *Die Assimilation der arabischen Medizin durch das lateinische Mittelalter.* Wiesbaden, 1964.

Schmitt, C. B. "Aristotle Among the Physicians." In *The Medical Renaissance of the Sixteenth Century.* Ed. A. Wear, et al. 1–15. Cambridge, 1985.

Schoener, E. "Das Viererschema in der antiken Humoralpathologie." *Sudhoff Archiv* 48, no. 4 (1964).

Schone, H. *Appolonios von Kitios.* Leipzig, 1896.

Schubring, K. "Bemerkungen zu der Galenausgabe von K. G. Kühn und zu ihrem Nachdruck." In Galen, *Opera omnia.* Vol. 20. Ed. K. G. Kühn. Hildesheim, 1965 (1st ed. Leipzig, 1822–31), pp. I–LXII.

Schultz, J. B. *Art and Anatomy in Renaissance Italy.* Ann Arbor, 1982.

———. "A Fifteenth-Century Papal Brief on Human Dissection." *Medical Heritage* 2 (1986): 50–56.

Schupbach, W. "The Paradox of Rembrandt's 'Anatomy of Dr. Tulp.'" *Medical History* (London), supp. 2 (1982).

Serlio, S. *Regole generali di architettura . . . sopra le cinque maniere degli edifici, cioè Toscano, Dorico, Jonico, Corintio e Composito con gli esempi delle antichità, che per la maggior parte concordano con la dottrina di Vitruvio.* Venice: F. Marcolini, 1537.

———. *Il secondo libro d'architettura.* Paris: Iehan Barbé, 1545.

Sigerist, H. E. *A History of Medicine.* 2 vols. Oxford, 1951–61.

Simeoni, L., and A. Sorbelli. *Storia dell'Università di Bologna.* Bologna, 1940.

Singer, C. "The Confluence of Humanism, Anatomy and Art." In *Fritz Saxl 1890–1948: A Volume of Memorial Essays from his Friends in England.* Ed. D. J. Gordon, 261–69. London, 1957.

———. *The Evolution of Anatomy: a Short History of Anatomical and Physiological Discovery to Harvey.* London, 1925.

———. *Galen on Anatomical Procedures.* London, 1956.

———. *A Short History of Anatomy and Physiology from the Greeks to Harvey.* New York, 1957.

———. "A Study on Early Renaissance Anatomy. With a New Text: the *Anathomia* of Hieronymo Manfredi (1490), Text Transcribed and Translated by Westland." In *Studies in the History and Method of Science.* Vol. 1. Ed. C. Singer, 79–164. Oxford, 1917–21.

Singer, C., and R. Rabin. *A Prelude to Modern Science: Being a Discussion of the History, Sources and Circumstances of the "Tabulae anatomicae sex" of Vesalius.* Cambridge, 1946.

Siraisi, N. *Medieval and Early Renaissance Medicine: An Introduction to Knowledge and Practice.* Chicago and London, 1990.

———. *Taddeo Aldarotti and his Pupils.* Princeton, 1981.

———. "Vesalius and Human Diversity in *De humani corporis fabrica.*" *Journal of the Warburg and Courtauld Institutes* 57 (1994): 60–88.

———. "Vesalius and the Reading of Galen's Teleology." *Renaissance Quarterly* 50 (1997): 1–37.

Spielmann, M. H. *The Iconography of Andreas Vesalius*. London, 1925.

Staden, H. von. "Anatomy as Rhetoric: Galen on Dissection and Persuasion." *Journal of the History of Medicine and Allied Sciences* 50 (1995): 47–66.

———. *Herophilus. The Art of Medicine in Early Alexandria*. Cambridge and London, 1989.

Steckerl, F. *The Fragments of Praxagoras of Cos and His School*. Leiden, 1958.

Steinberg, L. "Michelangelo and the Doctors." *Bulletin of the History of Medicine* 56 (1982): 543–53.

Steiner, F. B. *Taboo*. London, 1956.

Steinschneider, M. *Die arabischen Übersetzungen aus dem griechischen*. Graz, 1960.

———. "Constantinus Africanus und seine arabischen Quellen." *Archiv für pathologische Anatomie* 37 (1866): 351–410.

Sterzi, G. "Giulio Casserio, anatomico e chirurgo (1552–1616)." *Nuovo Archivio Veneto,* n. s. 9 (1909): 208–78; 10 (1910): 1–103.

Stroppiana, L. "L'anatomia nel Corpus Hippocraticum." *Rivista di Storia della Medicina* 7 (1963): 9–17.

Sudhoff, K. "Die vierte Salernitaner Anatomie." *Archiv für Geschichte der Medizin* 20 (1928).

Summers, D. *Michelangelo and the Language of Art*. Princeton, 1981.

Tabanelli, M., ed. *La chirurgia italiana nell'alto medioevo*. Florence, 1965.

Talbot, H. "Medicine." In *Science and the Middle Ages*. Ed. D.C. Lindberg. Chicago and London, 1978, 391–428.

Temkin, O. *Galenism. Rise and Decline of a Medical Philosophy*. Ithaca and London, 1973.

———. *Soranos' Gynecology*. Baltimore, 1956.

———. "Studies on Late Alexandrian Medicine. I. Alexandrinus' Commentaries on Galen's *De sectis ad Introducendos*." *Bulletin of the Institute of the History of Medicine* 3 (1935): 405–30.

———. "Was Servetus Influenced by Ibn al-Nafis?" *Bulletin of the History of Medicine* 12 (1940): 731–34.

Tenenti, A. *Il senso della morte e l'amore della vita nel Rinascimento*. Turin, 1978 (1st ed. Turin, 1957).

Tentler, T. N. *Sin and Confession on the Eve of the Reformation*. Princeton, 1977.

Tertullian. *De anima*. Ed. J. H. Waszink. Amsterdam, 1947.

Theophanes. *Chronographia*. Ed. C. de Boor. Leipzig, 1883.

Theophilus Protospartarius. *De corporis humani fabrica libri quinque*. . . . Venice [Heirs of O. Scotus], 1537.

Thomas of Aquinas. *Sancti Thomae Aquinatis Opera Omnia, Iussu Leonis XII P. M. Edita*. Rome, 1882–.

Thorndike, L. *History of Magic and Experimental Science*. 8 vols. New York, 1934–58.

———. "Translations of the Works of Galen from the Greek by Pietro d'Abano." *Isis* 30 (1942): 649–53.

Tietze-Conrat, E. "Neglected Contemporary Sources Relating to Michelangelo and Titian." *The Art Bulletin* 25 (1943): 156–59.

Tomasini, G. F. *Gymnasium Patavinum*. Udine, 1654.

Toni, G., and P. di Pietro. *L'Insegnamento dell' Anatomia nello Studio Bolognese.* Modena, 1971.

Tornikes, G. *Lettres et discours.* Ed. J. Darrouzès. Paris, 1970.

Tosoni, P. *Della anatomia degli antichi e della scuola anatomica padovana.* Padua, 1844.

Ullmann, M. *Die Medizin im Islam.* Leiden, 1970.

————. *Islamic Medicine.* Edinburgh, 1978 (1st ed. [German] Leiden, 1970).

Valverde de Hamusco, J. de. *Anatomia del corpo humano . . . con molte figure di rame ed eruditi discorsi in luce mandata.* Rome: A. Salamanca and A. Lafreri, 1560.

————. *La anatomia del corpo umano.* Venice: Stamperia de Giunti, 1586.

————. *Historia de la composicion del cuorpo humano.* Rome: A. Salamanca and A. Lafreri, 1556.

Vasari, G. *Le vite de' piú eccellenti pittori, scultori e architettori.* Ed. G. Milanesi. 9 vols. Florence, 1878–85. Reprint, Florence, 1981. [Based on Florence, 1568 ed.]

Vegetti, M. "L'animale ridicolo." In M. Vegetti. *Tra Edipo e Euclide: Forme del sapere antico.* Milan, 1983, 59–70.

————. *Il coltello e lo stilo.* Milan, 1987.

————. "Modelli di medicina in Galeno." In *Galen: Problems and Prospects.* Ed. V. Nutton. London, 1981.

————. "La scienza ellenistica: problemi di epistemologia storica." In M. Vegetti. *Tra Edipo e Euclide. Forme del sapere antico.* Milan, 1983, 151–91.

Vernet, J. *La cultura hispano-árabe en Oriente y Occidente.* Madrid, 1978.

Vesalius, A. *Epistola [ad Joachinum Roelants], rationem modumque propinandi radicis Chynae decocti . . . , pertractans; et praeter alia quaedam, epistolae cujusdam ad Jacobum Sylvium sententiam recensens, veritatis ac potissimum humanae fabricae studiosis perutilem; quum qui hactenus in illa nimium Galeno creditum sit, facile commonstret. Accessit quoque locuples rerum et verborum in hac ipsa epistola memorabilium index . . . [A. Francisco Vesalio, auctoris fratre, editum].* Basel: J. Oporinus, 1546.

————. *De humani corporis fabrica Epitome.* Basel: J. Oporinus, 1543.

————. *De humani corporis fabrica libri septem.* Basel: J. Oporinus, 1543. Reprint, Basel: J. Oporinus, 1555.

————. *Tabulae anatomicae sex.* Venice: B. Vitali, 1538.

————. *Von des menschen Cörpers Anatomey, ein kurtzer aber vast nützer Ausszug auss D. Andr. Vesalii . . . Bücheren von ihm selbs in Latein beschriben unnd durch D. Albanum Torinum verdolmetscht.* Basel: J. Oporinus, 1543.

Vialles, N. *Le sang et la chair. Les abatoirs du pays de l'Adour.* Paris, 1987.

Viano, C. A. "Lo scetticismo antico e la medicina." In *Lo scetticismo antico.* Ed. G. Giannantoni. Naples, 1981, 563–656.

Vindicianus. "Gynaecia." In *Teodori Prisciani Euporiston libri III.* Ed. V. Rose, 425–66. Leipzig, 1894.

Walsh, J. J. "The Popes and the History of Anatomy." *Medical Library and History Journal* 2 (1904): 10–28.

————. *The Popes and Science.* New York, 1911.

Wells, L. H. "A remarkable Pair of Anatomical Fugitive Sheets." *Bulletin of the History of Medicine* 37 (1964): 470–76.

Wickersheimer, E. "L'anatomie de Guido de Vigevano (1345)." *Archiv für Geschichte der Medizin* 7 (1914): 1–25.

———. *Anatomies de Mondino dei Liuzzi et de Guido de Vigevano*. Paris, 1926.

———. "Les premières dissections à la faculté de médecine de Paris." *Bulletin de la société d'histoire de Paris et de l'Ile-de-France* 37 (1910): 159–69.

———., ed. *Commentaires de la Faculté de Médecine de Paris (1395–1516)*. Paris, 1915.

Wilson, L. "William Harvey's 'Praelectiones:' The Performance of the Body in the Renaissance Theater of Anatomy." *Representations* 17 (1987): 62–95.

Winther, J. (Iohannes Guinterius Andernacus). *Institutionum Anatomicarum secundum Galeni sententiam ad candidatos Medicinae libri quatuor*. Basel: B. Lasium and T. Platter, 1536.

Wissowa, G., et al., eds. *Pauly's Real-Encyclopädie der Klassischen Altertumwissenschaft*. Stuttgart, 1893–1978.

White, A. D. *The Warfare for Science*. New York, 1876. (Italian trans.: *Storia della lotta della scienza con la teologia nella cristianità*. Turin, 1902.)

Wolf-Heiddeger, G., and A. M. Cetto. *Die anatomische Sektion in bildlicher Darstellung*. Basel and New York, 1967.

Zerbi, G. *Liber Anathomiae corporis humani et singulorum memborum*. Venice: B. Locatellum, 1502.

INDEX

Accademia del Disegno, 65fig.
Achillini, Alessandro, 197, 198
Aegineta, Paulus, 28fig., 30, 30n. 39
Aesculapius, 28fig., 29fig., 52n. 80,
  142n. 66
Aetius, 150
Albucasis, 150
Alcmaeon of Croton, 123, 124n. 6,
  130, 131
Alderotti, Taddeo, 10n. 5
Alexander of Aphrodisias, 198n. 29
Alexander of Hales, 111n. 136
Alexander of Tralles, 150
Alexandria: Alexandrian Dogmatist
  school and, 142, 147; dissection in,
  121, 121n. 1, 129, 138–39, 139n.
  139, 143, 149, 156, 162, 162n.
  132, 167; embalming practices and,
  161–62; Galen in, 143; Herophilus
  and, 134–35, 135n. 45, 46, 136,
  136n. 49, 167; Hippocratic medi-
  cine in, 124n. 6, 134, 134n. 44;
  Numisianus in, 140
Alexandrinus, Johannes, 198, 198n.
  30
Ambrose, Saint, 161n. 127
Analytica Posteriora, 72n. 9
Anathomia (Manfredi), 197
Anathomia Richardi, 181–82, 183
Anatomia (Mondino dei Liuzzi), 181,

189; 1493 original anatomy lesson
  illustration of, 10–11, 12fig., 13,
  14; 1495 Leipzig edition, 16–18,
  16–17n. 17, 19, 19fig.; 1514 Rostock edition, 14; 1519 Geneva edition, 15–16, 17fig.; 1522 Venice edition, 14, 14n. 12,, 15fig.; autopsy
  and, 179n. 172; Berengario cri-
  tiqued by, 21, 22n. 24, 23fig.,
  26fig.; Corti and, 205; dissection in,
  170n. 151, 183; Galenic character
  of, 10, 11, 19, 151, 170–72, 171nn.
  152, 153, 155, 176, 195–96, 197;
  inconsistencies resolved in, 19–20;
  Massa and, 199; popularity of, 2,
  11, 20–21, 24, 85–86, 195–96, 206;
  reprintings of, 196, 196n. 25; text
  vs. illustration discrepancy and, 25
Anatomia del corpo umano
  (Valverde), 95n. 84; 1560 engraving
  and, 56; 1586 Venetian reprinting
  of, 56, 57fig., 58fig.; Roman edition, 54, 54n. 85, 55fig., 56, 56n.
  86; unchanged title pages and, 56,
  56n. 87
Anatomia matricis (Galen), 195
Anatomia oculi (Galen), 195
*Anatomia porci ex traditione Co-
  phonis* (Dryander), 144n. 79, 154n.
  107, 223–24, 223nn. 91, 92

251

dissection, text authority, questioning
of: Du Bois and, 207–13, 208nn.
58, 59, 209nn. 64, 65, 67, 222,
225; Eustachio and, 211–12, 212n.
77; *rete mirabile* issue and, 201–2,
203fig.; by Vesalius, 202–3, 202n.
46, 203fig., 206–7, 207nn. 54, 55,
228; Vesalius's errors and, 206,
206n. 51, 207–8; Vesalius vs. Corti
and, 204–6; Vesalius vs. Du Bois
and, 207–8, 207–8nn. 57–59
dissection, text authority vs. observation and, 187; Achillini's *Annotationes* and, 197, 198; Berengario's
Commentaria and, 198–99, 199n.
33, 212; Catholic vs. Protestant conflict and, 212–13; Galenic anatomy
focus and, 195–98, 195nn. 19, 20,
197n. 29; Galen's errors and, 198–
201, 198n. 30, 199n. 36, 199nn.
33, 36, 201, 201n. 41; Massa's
Liber introductorius anatomicae
and, 199–201, 200–201nn. 37–41,
212; Mondino's *Anatomia* and,
195–96, 196n. 25, 197; Vesalius
and, 185, 215; writers' attitudes
and, 194–95
dissection scene, iconographic investigation of: *The Anatomical Table of
John Banister* (Hilliard) and, 57–59,
59fig.; *De corporis humani structura et usu . . .* (Platter) and, 66fig.;
De humani corporis fabrica epitome
(Vesalius) and, 39, 39n. 49, 42,
42n. 57, 45fig., 47n. 70, 52n. 79,
57n. 69, 60, 188n. 1, 205; *El libro
de proprietatibus rerum* (Bartholomaeus Anglicus) and, 35, 36, 37fig.;
*Historia de la composicion del cu-
orpo humano* (Valverde) and, 53–
54, 54n. 85; illustrated title page
and, 8–9; *Le livre appelé Guidon*
(Chauliac) and, 35, 35n. 48; *Tabulae anatomicae sex* (Vesalius) and,
42, 42n. 58, 47n. 69; Vesalius influence upon, 53–68, 53n. 82, 54n.
85, 55fig., 56nn. 86, 87, 60fig.,
64n. 107. *See also Anatomia* (Mon-

dino); *Anatomia del corpo umano*
(Valverde); *Commentaria* (Berengario); *De anatomicis administrationibus* (Galen); *De humani corporis fabrica libri septem* (Vesalius);
De re anatomica libri XV (Colombo); *Fasciculus medicinae* (John
of Ketham); *Le propriétaire des
choses* (Bartholomaeus Anglicus)
Dogmatist physicians: Alexandrian
Dogmatist school and, 142, 147;
anatomical knowledge and, 122,
122n. 2, 136–38, 139n. 57, 149,
162n. 131, 169, 180
Dou, Gerald, 64n. 107, 67fig.
Dryander, J., 144n. 79, 154n. 107,
223–24, 223nn. 91, 92, 225
Du Bois, Jacques (Iacobus Sylvius),
51, 207–13, 208nn. 58, 59, 209nn.
64, 65, 67, 222, 225
Du nom des parties (Rufus), 162n.
132
Dürer, Albrecht, 65, 65n. 110
Dyamantier, Jean Jenin le, 35n. 45

Edelstein, Ludwig, 4, 121n. 1, 126,
128, 139n. 57, 148n. 90
El libro de proprietatibus rerum (Bartholomaeus Anglicus), 35, 36, 37fig.
Embalming, mummification practices,
161–62
Empiricist physicians: anatomical
knowledge and, 122, 122n. 2, 123,
136–38, 139n. 57, 162n. 131, 169;
contamination from body and, 162;
dissection, vivisection and, 129n.
121, 158–59, 159nn. 121, 122,
124, 162, 163, 166, 169; experiential focus of, 157–60, 158nn. 117–
120, 160n. 126; rational discipline
focus of, 141; wound investigation
and, 162, 170
Epistle to the Galatians (St. Paul),
113, 113n. 141
Erasistratus: anatomical knowledge
and, 124n. 7, 206; dissection and,
135–36, 135n. 45, 137, 158, 215n.
81; Galen and, 143; origin of birth